Status of Groundwater Quality in the San Fernando–San Gabriel Study Unit, 2005: California GAMA Priority Basin Project

By Michael Land, Justin T. Kulongoski, and Kenneth Belitz

A product of the California Groundwater Ambient Monitoring and Assessment (GAMA) Program

Prepared in cooperation with the California State Water Resources Control Board

Scientific Investigations Report 2011–5206

U.S. Department of the Interior
U.S. Geological Survey

U.S. Department of the Interior
KEN SALAZAR, Secretary

U.S. Geological Survey
Marcia K. McNutt, Director

U.S. Geological Survey, Reston, Virginia: 2012

For more information on the USGS—the Federal source for science about the Earth, its natural and living resources, natural hazards, and the environment, visit http://www.usgs.gov or call 1–888–ASK–USGS.

For an overview of USGS information products, including maps, imagery, and publications, visit http://www.usgs.gov/pubprod

To order this and other USGS information products, visit http://store.usgs.gov

Suggested citation:
Land, Michael, Kulongoski, J.T., and Belitz, Kenneth, 2012, Status of groundwater quality in the San Fernando—San Gabriel study unit, 2005—California GAMA Priority Basin Project: U.S. Geological Survey Scientific Investigations Report 2011–5206, 66 p.

Contents

Figures

Figures—Continued

Tables

Conversion Factors, Datums, and Abbreviations and Acronyms

Conversion Factors

Inch/foot/mile to International System of Units (SI)

Multiply	By	To obtain
Length		
inch (in.)	2.54	centimeter (cm)
inch (in.)	25.4	millimeter (mm)
foot (ft)	0.3048	meter (m)
mile (mi)	1.609	kilometer (km)
Area		
square foot (ft^2)	0.09290	square meter (m^2)
square mile (mi^2)	2.590	square kilometer (km^2)
Flow rate		
acre-foot per year (acre-ft/yr)	1,233	cubic meter per year (m^3/yr)
inch per year (in/yr)	25.4	millimeter per year (mm/yr)
Radioactivity		
picocurie per liter (pCi/L)	0.037	becquerel per liter (Bq/L)

Temperature in degrees Celsius (°C) may be converted to degrees Fahrenheit (°F) as follows:

$$°F=(1.8×°C)+32.$$

Temperature in degrees Fahrenheit (°F) may be converted to degrees Celsius (°C) as follows:

$$°C=(°F−32)/1.8.$$

Specific conductance is given in microsiemens per centimeter at 25 degrees Celsius (µS/cm at 25 °C).

Concentrations of chemical constituents in water are given either in milligrams per liter (mg/L) or micrograms per liter (µg/L) or nanograms per liter (ng/L). One milligram per liter is equivalent to 1 part per million (ppm); 1 microgram per liter is equivalent to 1 part per billion (ppb); 1 nanogram per liter (ng/L) is equivalent to 1 part per trillion (ppt); 1 per mil is equivalent to 1 part per thousand.

Datums

Vertical coordinate information is referenced to the North American Vertical Datum of 1988 (NAVD 88).

Horizontal coordinate information is referenced to the North American Datum of 1983 (NAD 83).

Conversion Factors, Datums, and Abbreviations and Acronyms—Continued

Abbreviations and Acronyms

AL-US	U.S. Environmental Protection Agency action level
BLS	below land surface
FG	San Fernando–San Gabriel GAMA Priority Basin Project study unit
GAMA	Groundwater Ambient Monitoring and Assessment Program
HAL-US	U.S. Environmental Protection Agency lifetime health advisory level
HBSL	health-based screening level
HTO	tritiated water
LRL	laboratory reporting level
LSD	land-surface datum
MCL	maximum contaminant level
MCL-CA	California Department of Public Health maximum contaminant level
MCL-US	U.S. Environmental Protection Agency maximum contaminant level
MDL	method detection level
MRL	minimum reporting level
NL-CA	California Department of Public Health notification level
pmc	percent modern carbon
RSD5-US	U.S. Environmental Protection Agency risk-specific dose at a risk factor of 10^{-5}
SMCL	secondary maximum contaminant level
SMCL-CA	California Department of Public Health secondary maximum contaminant level
SMCL-US	U.S. Environmental Protection Agency secondary maximum contaminant level
SF	San Fernando study area
SG	San Gabriel study area
TEAP	terminal electron acceptor processes
TT-US	U.S. Environmental Protection Agency treatment technique levels

Organizations

CDPH	California Department of Public Health (Department of Health Services prior to July 1, 2007)
CDPR	California Department of Pesticide Regulation
CDWR	California Department of Water Resources
LLNL	Lawrence Livermore National Laboratory
SWRCB	State Water Resources Control Board (California)
USEPA	U.S. Environmental Protection Agency
USGS	U.S. Geological Survey

Conversion Factors, Datums, and Abbreviations and Acronyms—Continued

Selected Chemical Names

CFC	chlorofluorocarbon
DBCP	1,2-dibromo-3-chloropropane
DEHP	bis(2-ethylhexyl) phthalate
DOC	dissolved organic carbon
EDB	1,2-dibromomethane (ethylene dibromide)
MTBE	methyl *tert*-butyl ether
NDMA	*N*-nitrosodimethylamine
Nitrate-N	nitrate as nitrogen
Nitrite-N	nitrite as nitrogen
PCE	tetrachloroethene or perchloroethene
TCE	trichloroethene
1,2,3-TCP	1,2,3-trichloropropane
TDS	total dissolved solids
THM	trihalomethane
VOC	volatile organic compound
Ammonia-N	ammonia as nitrogen

Units of Measure

a	annum (year)
cm^3 STP g^{-1}	cubic centimeters at standard temperature and pressure per gram
δ	delta notation; the ratio of a heavier isotope to the more common lighter isotope of an element, relative to a standard reference material, expressed as per mil
>	greater than
%	percent

Status of Groundwater Quality in the San Fernando–San Gabriel Study Unit, 2005: California GAMA Priority Basin Project

By Michael Land, Justin T. Kulongoski, and Kenneth Belitz

Abstract

Groundwater quality in the approximately 460-square-mile San Fernando–San Gabriel (FG) study unit was investigated as part of the Priority Basin Project of the Groundwater Ambient Monitoring and Assessment (GAMA) Program. The study area is in Los Angeles County and includes Tertiary-Quaternary sedimentary basins situated within the Transverse Ranges of southern California. The GAMA Priority Basin Project is being conducted by the California State Water Resources Control Board in collaboration with the U.S. Geological Survey (USGS) and the Lawrence Livermore National Laboratory.

The GAMA FG study was designed to provide a spatially unbiased assessment of the quality of untreated (raw) groundwater in the primary aquifer systems (hereinafter referred to as primary aquifers) throughout California. The assessment is based on water-quality and ancillary data collected in 2005 by the USGS from 35 wells and on water-quality data from the California Department of Public Health (CDPH) database. The primary aquifers were defined by the depth interval of the wells listed in the CDPH database for the FG study unit. The quality of groundwater in primary aquifers may be different from that in the shallower or deeper water-bearing zones; shallow groundwater may be more vulnerable to surficial contamination.

This study assesses the status of the current quality of the groundwater resource by using data from samples analyzed for volatile organic compounds (VOCs), pesticides, and naturally occurring inorganic constituents, such as major ions and trace elements. This *status assessment* is intended to characterize the quality of groundwater resources in the primary aquifers of the FG study unit, not the treated drinking water delivered to consumers by water purveyors.

Relative-concentrations (sample concentration divided by the health- or aesthetic-based benchmark concentration) were used for evaluating groundwater quality for those constituents that have Federal and (or) California regulatory or non-regulatory benchmarks for drinking-water quality. A relative-concentration greater than ($>$) 1.0 indicates a concentration greater than a benchmark, and less than or equal to ($\leq$) 1.0 indicates a concentration equal to or less than a benchmark. Relative-concentrations of organic and special-interest constituents [perchlorate, N-nitrosodimethylamine (NDMA), 1,4-dioxane, and 1,2,3-trichloropropane (1,2,3-TCP)] were classified as "high" (relative-concentration >1.0), "moderate" ($0.1<$ relative-concentration ≤ 1.0), or "low" (relative-concentration ≤ 0.1). Relative-concentrations of inorganic constituents were classified as "high" (relative-concentration > 1.0), "moderate" ($0.5 <$ relative-concentration ≤ 1.0), or "low" (relative-concentration ≤ 0.5).

Aquifer-scale proportion was used as the primary metric in the *status assessment* for evaluating regional-scale groundwater quality. High aquifer-scale proportion is defined as the percentage of the area of the primary aquifers with a relative-concentration greater than 1.0 for a particular constituent or class of constituents; percentage is based on an areal rather than a volumetric basis. Moderate and low aquifer-scale proportions were defined as the percentage of the primary aquifers with moderate and low relative-concentrations, respectively. Two statistical approaches—grid-based and spatially weighted—were used to evaluate aquifer-scale proportions for individual constituents and classes of constituents. Grid-based and spatially weighted estimates were comparable in the FG study unit (within 90-percent confidence intervals).

Inorganic constituents with human-health benchmarks were detected at high relative-concentrations in 9.1 percent of the primary aquifers and moderate in 33.3 percent. High aquifer-scale proportion of inorganic constituents primarily reflected high aquifer-scale proportions of nitrate (8.8 percent). The inorganic constituents with secondary maximum contaminant levels (SMCLs), iron, sulfate, and total dissolved solids (TDS) had relative-concentrations that were high in 3.2 percent and moderate in 18.2 percent of the primary aquifers.

Relative-concentrations of organic constituents (one or more) were high in 18.2 percent, and moderate in 42.9 percent, of the primary aquifers, based on the spatially weighted approach. The high aquifer-scale proportion of organic constituents primarily reflected high aquifer-scale proportions of trichloroethene (TCE; 14.8 percent), perchloroethene

(PCE; 11.2 percent), and carbon tetrachloride (6.5 percent). Of the 212 organic and special-interest constituents analyzed, 66 constituents were detected. Chloroform, PCE, simazine, atrazine, and TCE were each detected in more than 50 percent of the 35 grid wells. Bromodichloromethane, *cis*-1,2-dichloroethene, 1,1-dichloroethane, perchlorate, carbon tetrachloride, and 1,1-dichloroethene were detected in more than 30 percent of the grid wells. Methyl *tert*-butyl ether (MTBE), prometon, and diuron were detected in more than 20 percent of the grid wells, and CFC-12, bromacil, carbon disulfide, 1,1,1-trichloroethane (TCA), CFC-113, tebuthiuron, dibromochloromethane, and CFC-11 were detected in more than 10 percent of the grid wells. However, perchlorate, diuron, and bromacil were sampled only in a subset of 11 wells, not in all 35 grid wells. Perchlorate and NDMA were detected at high relative-concentrations in 11.2 percent and 5.2 percent of the primary aquifers, respectively, based on the spatially weighted approach. Pharmaceutical compounds were not detected at concentrations greater than or equal to method detection limits in the study unit.

Introduction

To assess the quality of ambient groundwater in aquifers used for drinking-water supply and to establish a baseline groundwater-quality monitoring program, the State Water Resources Control Board (SWRCB), in collaboration with the U.S. Geological Survey (USGS) and Lawrence Livermore National Laboratory (LLNL), implemented the Groundwater Ambient Monitoring and Assessment (GAMA) Program (California Environmental Protection Agency, 2010, website at http://www.waterboards.ca.gov/gama/). The statewide GAMA Program currently consists of three projects: (1) the GAMA Priority Basin Project, conducted by the USGS (U.S. Geological Survey, 2010, website at http://ca.water.usgs.gov/gama/); (2) the GAMA Domestic Well Project, conducted by the SWRCB; and (3) the GAMA Special Studies, conducted by LLNL. On a statewide basis, the Priority Basin Project focused on the primary aquifers, typically the deep portion of the groundwater resource, and the SWRCB Domestic Well Project generally focused on the shallow aquifer systems. The deeper aquifers may be at less risk of contamination than the shallow wells, such as private domestic and environmental monitoring wells, which are closer to surficial sources of contamination. As a result, concentrations of contaminants, such as volatile organic compounds (VOCs) and nitrate, in wells screened in the deep aquifers may be lower than concentrations of constituents in shallow wells (Kulongoski and others, 2010; Landon and others, 2010).

The SWRCB initiated the GAMA Program in 2000 in response to Legislative mandates (State of California, 1999, 2001a, Supplemental Report of the 1999 Budget Act 1999–00 Fiscal Year). The GAMA Priority Basin Project was initiated

in response to the Groundwater Quality Monitoring Act of 2001 (State of California, 2001b, Sections 10780–10782.3 of the California Water Code, Assembly Bill 599) to assess and monitor the quality of groundwater in California. The GAMA Priority Basin Project is a comprehensive assessment of statewide groundwater quality designed to help better understand and identify risks to groundwater resources and to increase the availability of information about groundwater quality to the public. For the Priority Basin Project, the USGS, in collaboration with the SWRCB, developed a monitoring plan to assess groundwater basins through direct sampling of groundwater and other statistically reliable sampling approaches (Belitz and others, 2003; California State Water Resources Control Board, 2003). Additional partners in the GAMA Priority Basin Project include the California Department of Public Health (CDPH), the California Department of Pesticide Regulation (CDPR), the California Department of Water Resources (CDWR), and local water agencies and well owners (Kulongoski and Belitz, 2004).

The range of hydrologic, geologic, and climatic conditions that exist in California must be considered in an assessment of groundwater quality. Belitz and others (2003) partitioned the State into 10 hydrogeologic provinces, each with distinctive hydrologic, geologic, and climatic characteristics (fig. 1). All these hydrogeologic provinces include groundwater basins and subbasins designated by the CDWR (California Department of Water Resources, 2003). Groundwater basins generally consist of relatively permeable, unconsolidated deposits of alluvial or volcanic origin. Eighty percent of California's approximately 16,000 public-supply wells are in designated groundwater basins. Groundwater basins and subbasins were prioritized for sampling on the basis of the number of public-supply wells, with secondary consideration given to municipal groundwater use, agricultural pumping, the number of historically leaking underground fuel tanks, and registered pesticide applications (Belitz and others, 2003). The 116 priority basins and additional areas outside defined groundwater basins were grouped into 35 study units, which include approximately 95 percent of public-supply wells in California's groundwater basins.

Purpose and Scope

The purposes of this report are to provide a (1) *study unit description*: description of the hydrogeologic setting of the San Fernando–San Gabriel study unit (fig. 1), hereinafter referred to as the FG study unit, and (2) *status assessment*: assessment of the status of the current quality of groundwater in the primary aquifers in the FG study unit.

Water-quality data for samples collected by the USGS for the GAMA Program in the FG study unit and details of sample collection, analysis, and quality-assurance procedures for the FG study unit are reported by Land and Belitz (2008).

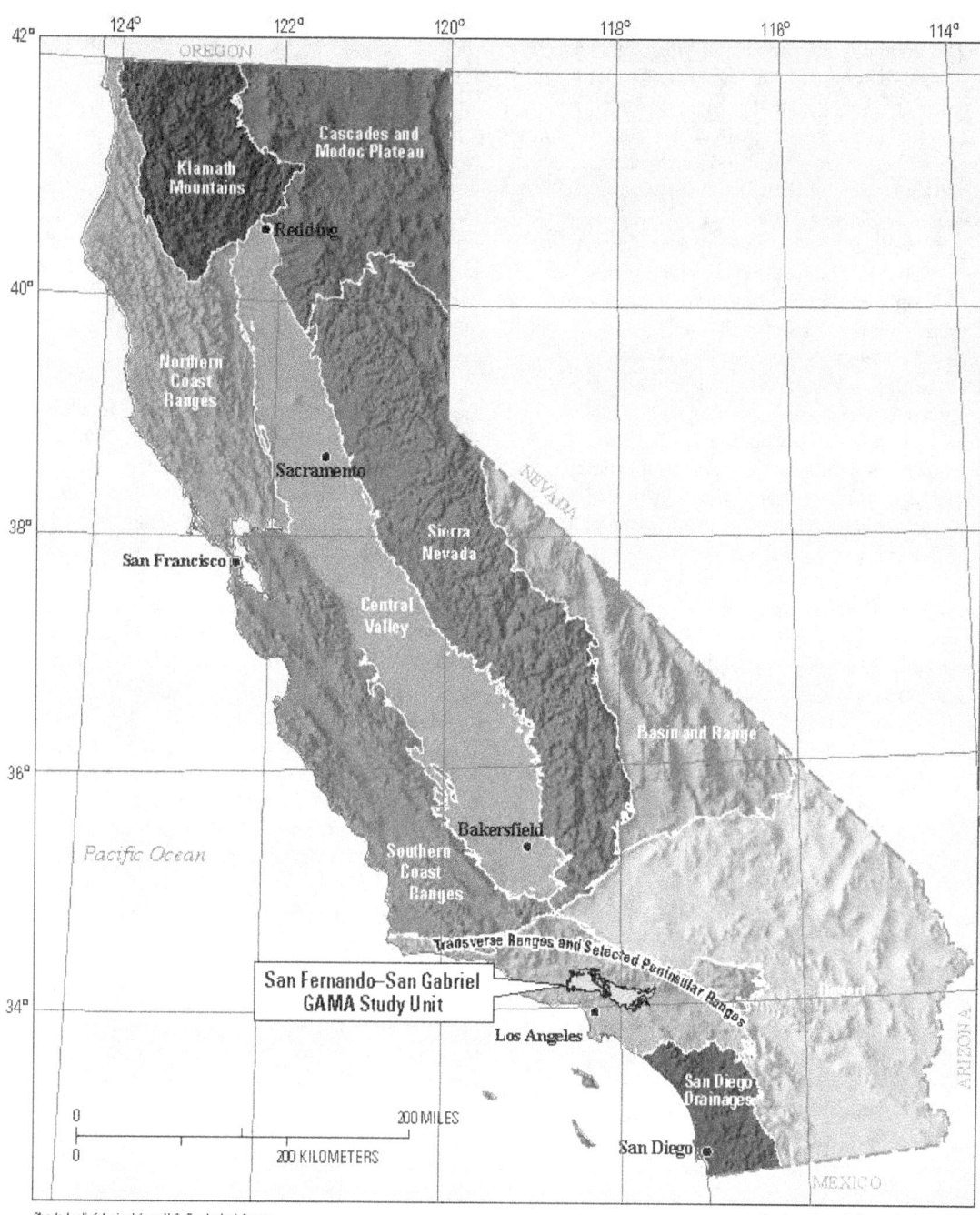

Shaded relief derived from U.S. Geological Survey
National Elevation Dataset, 2006,
Albers Equal Area Conic Projection

Provinces from Belitz and others, 2003

Figure 1.   Location of the San Fernando–San Gabriel study unit, California GAMA Priority Basin Project, and the California hydrogeologic provinces.

Utilizing those same data, this report describes methods used in designing the sampling network, identifying CDPH data for use in the status assessment, estimating aquifer-scale proportions of relative-concentrations, analyzing ancillary datasets, classifying groundwater age, and assessing the status of groundwater quality by statistical and graphical approaches.

The status assessment includes analyses of water-quality data for 35 wells selected by the USGS for spatial coverage of one well per grid cell (hereinafter referred to as USGS-grid wells) across the FG study unit. Most of the USGS-grid wells were public-supply wells, but a few industrial and irrigation wells with perforation depth intervals similar to the public-supply wells also were sampled. Samples were collected for analysis of anthropogenic constituents, such as VOCs and pesticides, and naturally occurring inorganic constituents, such as major ions and trace elements. Water-quality data from the CDPH database also were used to supplement data collected by the USGS for the GAMA Program. The resulting set of water-quality data from USGS-grid wells and selected CDPH wells was considered to be representative of the primary aquifer systems (hereinafter referred to as primary aquifers) in the FG study unit; the primary aquifers are defined by the depth of the screened/perforated intervals of the wells listed in the CDPH database for the FG study unit. GAMA status assessments are designed to provide a statistically robust characterization of groundwater quality in the primary aquifers at the basin-scale (Belitz and others, 2003). The statistically robust design also allows basins to be compared, and results to be synthesized, regionally and statewide.

To provide context, the water-quality data discussed in this report are compared to California and Federal regulatory and non-regulatory benchmarks for drinking water. The assessments in this report are intended to characterize the quality of untreated groundwater resources in the primary aquifers within the study unit, not the drinking water delivered to consumers by water purveyors. This study does not attempt to evaluate the quality of water delivered to consumers; after withdrawal from the ground, water typically is treated, disinfected, and (or) blended with other waters to maintain acceptable water quality. Regulatory benchmarks apply to drinking water that is delivered to the consumer, not to untreated groundwater.

Hydrogeology of the San Fernando– San Gabriel Study Unit

The FG GAMA study unit covers approximately 460 square miles (mi²) in Los Angeles County, and includes a population of nearly 4 million people. The FG study unit lies within the Transverse Ranges and selected Peninsular Ranges hydrogeologic province (fig. 1) (Belitz and others, 2003)

and includes three groundwater basins (fig. 2): San Fernando Valley, Raymond, and San Gabriel Valley. For the purpose of this study, these three groundwater basins were grouped into two study areas, based primarily on location. The western part of the FG study unit includes the San Fernando Valley groundwater basin and is referred to as the San Fernando study area. The eastern part of FG study unit includes the Raymond and San Gabriel Valley groundwater basins, both of which are within the San Gabriel Valley, and collectively are referred to as the San Gabriel study area (fig. 2). As part of the Priority Basin Project, untreated groundwater samples were collected from 52 wells in the FG study unit during May 24–July 20, 2005.

The San Fernando and San Gabriel Valleys are sedimentologically diverse basins situated within the Transverse Ranges of southern California. These structurally complex basins formed as a result of the dextral slip of the San Andreas Fault system in the late Tertiary-Quaternary (Tinsley, 2001). The geological structure of the Transverse Ranges is dominated by the effects of north-south compressive deformation resulting in thrust faulting, strike-slip faulting, and bedrock folding. These deformations are attributable to convergence between the bend of the San Andreas Fault and northwestern motion of the Pacific Plate and are expressed in thrust faulting such as that exhibited by the 1994 Northridge earthquake, the 1971 San Fernando earthquake, and the 1987 Whittier Narrows earthquake (U.S. Geological Survey, 1996).

The FG study unit is bounded to the north by the Santa Susana and the San Gabriel Mountains, to the east by the San Jose and the Chino Faults (fig. 3), and to the south by the Santa Monica Mountains and the Elysian, Repetto, Merced and Puente Hills (fig. 2). The FG study unit has approximately 3,000 feet (ft) of topographic relief; the land surface of the groundwater basins slopes gently, with median altitudes ranging from approximately 600 to 1,000 ft above sea level.

The San Fernando Valley contains the headwaters of the Los Angeles River and its tributaries, and includes parts of the Bull Canyon, Sylmar, Tujunga, Verdugo, and Eagle Rock watersheds. The San Fernando Valley floor is composed of alluvial fan deposits that may be distinguished by origin as the western or eastern parts (Tinsley, 2001). East of Burbank, streams that emerge from Pacoima and Big Tujunga canyons drain the western San Gabriel Mountains and deposit coarse alluvium, while the shallower deposits in the western part of the valley derive mainly from Tertiary and pre-Tertiary sedimentary rocks, and are underlain by fine-grained deposits that are less permeable (Tinsley, 2001).

The San Gabriel (SG) study area includes parts of the Monk Hill, Pasadena, Lower Canyon, Upper San Gabriel River, Foothill, Live Oak, Pomona, and San Jose watersheds. The San Gabriel basin is filled primarily with alluvium deposited by streams flowing out of the San Gabriel Mountains. The deposits include Pleistocene and Holocene alluvium and the lower Pleistocene San Pedro Formation.

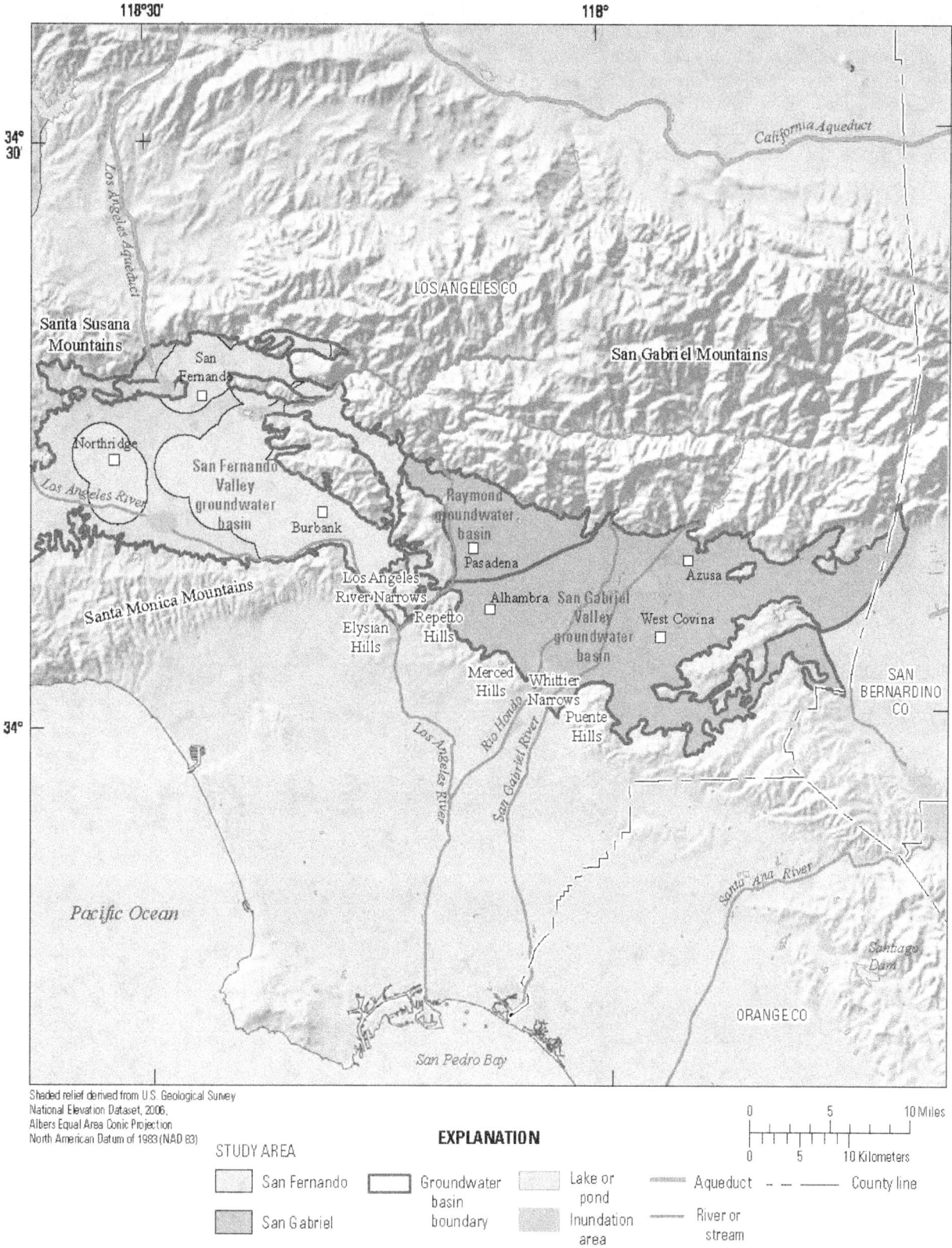

Figure 2. Geographic features and study areas of the San Fernando–San Gabriel study unit, California GAMA Priority Basin Project.

The alluvial fans along the San Gabriel Mountains and stream deposits following the course of the major streams across the valley consist primarily of highly permeable gravels and cobbles, with numerous interbedded lenses of clays also occurring, particularly in the southern portion of the basin and near the surrounding hills. The San Pedro Formation consists of interbedded marine sand, gravel, and silt grading eastward into continental alluvium (California Department of Water Resources, 2005a,b,c). In the San Gabriel Valley, the Raymond groundwater basin is separated from the San Gabriel Valley groundwater basin by the Raymond fault, which acts as a barrier to groundwater movement (fig. 3).

Figure 3. Geologic formations and areal distribution of USGS-grid and understanding wells sampled in the San Fernando–San Gabriel study unit, California GAMA Priority Basin Project.

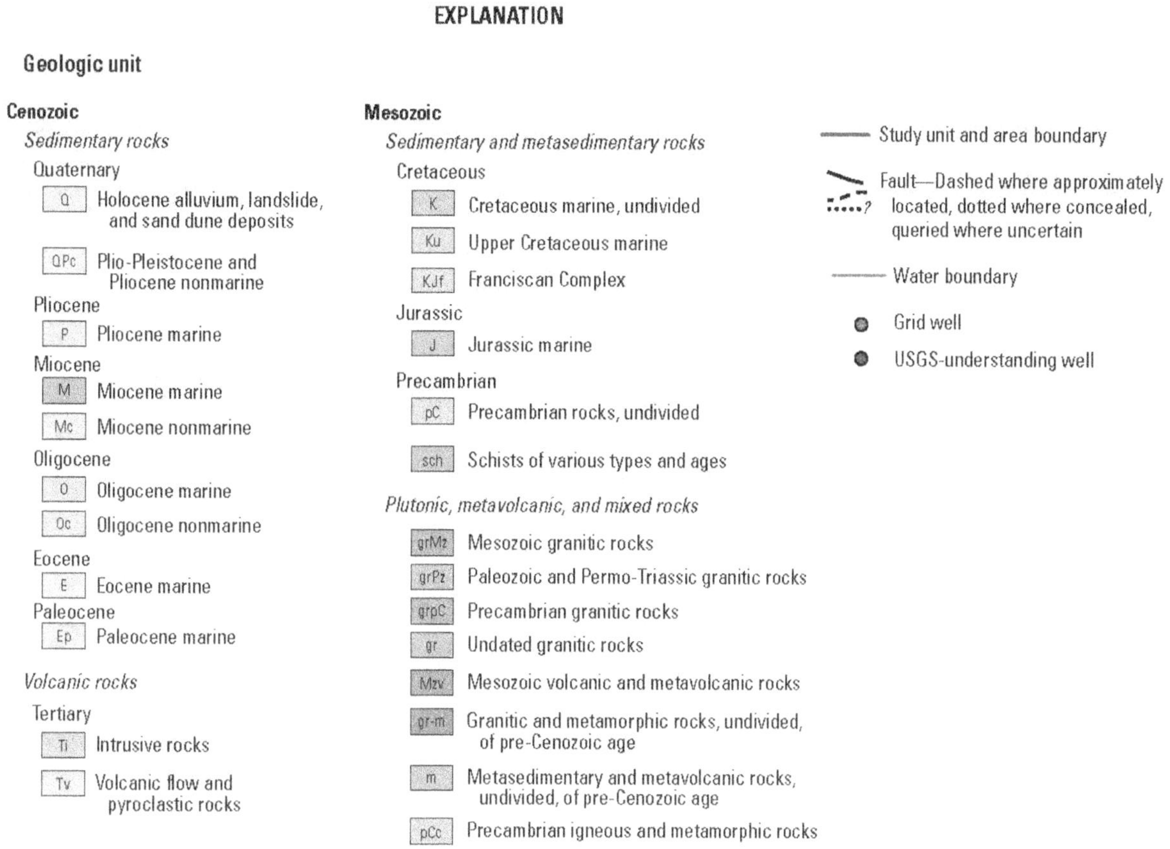

EXPLANATION

Geologic unit

Cenozoic

Sedimentary rocks

Quaternary

| Q | Holocene alluvium, landslide, and sand dune deposits

| QPc | Plio-Pleistocene and Pliocene nonmarine

Pliocene

| P | Pliocene marine

Miocene

| M | Miocene marine

| Mc | Miocene nonmarine

Oligocene

| O | Oligocene marine

| Oc | Oligocene nonmarine

Eocene

| E | Eocene marine

Paleocene

| Ep | Paleocene marine

Volcanic rocks

Tertiary

| Ti | Intrusive rocks

| Tv | Volcanic flow and pyroclastic rocks

Mesozoic

Sedimentary and metasedimentary rocks

Cretaceous

| K | Cretaceous marine, undivided

| Ku | Upper Cretaceous marine

| KJf | Franciscan Complex

Jurassic

| J | Jurassic marine

Precambrian

| pC | Precambrian rocks, undivided

| sch | Schists of various types and ages

Plutonic, metavolcanic, and mixed rocks

| grMz | Mesozoic granitic rocks

| grPz | Paleozoic and Permo-Triassic granitic rocks

| grpC | Precambrian granitic rocks

| gr | Undated granitic rocks

| Mzv | Mesozoic volcanic and metavolcanic rocks

| gr-m | Granitic and metamorphic rocks, undivided, of pre-Cenozoic age

| m | Metasedimentary and metavolcanic rocks, undivided, of pre-Cenozoic age

| pCc | Precambrian igneous and metamorphic rocks

——— Study unit and area boundary

⌇⌇⌇? Fault—Dashed where approximately located, dotted where concealed, queried where uncertain

——— Water boundary

◉ Grid well

● USGS-understanding well

Figure 3.—Continued.

The primary water-bearing materials in the FG study unit are composed of unconsolidated to semi-consolidated gravel, sand, and clay of lower Pleistocene to recent age (California Department of Water Resources, 2005a,b,c). The Holocene alluvium, up to 100 ft thick in the SG basin, 150 ft thick in the Raymond basin, and 900 ft thick near Burbank in the SF basin, consists of coarse-grained unsorted gravel and sand deposited by coalescing alluvial fans emanating from the surrounding highlands. Deeper water-bearing units consist of Pleistocene alluvial fan deposits, which are up to 4,100 ft thick in central portion of the SG basin, 1,140 ft thick in the Raymond basin, and unknown thickness in the SF basin, and lower Pleistocene marine deposits: the San Pedro Formation (sand, gravel, and silt), up to 2,000 ft thick in the San Gabriel basin, and the Saugus Formation (conglomerates, sands, silts, and clays), up to 6,400 ft thick in the central part of the San Fernando basin (California Department of Water Resources, 2005a,b,c). The lower Pleistocene marine deposits overlie crystalline basement. The primary aquifers targeted by this study include groundwater-bearing zones in which public-supply wells are completed. Public-supply wells are typically drilled to depths of 400 to 785 ft, consist of solid casing from the land surface to a depth of about 160 to 300 ft, and are perforated below the solid casing. Supply wells vary in depth depending on their location and depth of the alluvium.

The climate in the FG study unit is characterized by hot, dry summers and cool, moist winters, with most precipitation falling between the months of December and March (California Department of Water Resources, 2003).

Several creeks and washes drain the FG study unit. In the western part, water from surface channels drains to the Los Angeles River, where it passes through the Los Angeles River Narrows and into the coastal plain before ultimately reaching San Pedro Bay. In the eastern part, water from tributary creeks and washes drains to the San Gabriel River and the Rio Hondo, then passes through the Whittier Narrows before ultimately reaching San Pedro Bay. In the study areas, groundwater flow generally follows the topography of the basins, from high elevations towards the drainages, and down valleys towards the Pacific Ocean.

Recharge in the FG study unit is from a variety of sources. Recharge mainly is from direct infiltration of precipitation and irrigation, and infiltration of streamflow from the major rivers and their tributaries. Precipitation in FG study unit may vary from 15 inches per year (in/yr) in the valley areas to 31 in/yr in the upland areas; the average value over the three groundwater basins is 18 in/yr (California Department of Water Resources, 2005a,b,c). Runoff—consisting of natural streamflow, imported water, reclaimed wastewater, industrial discharge, and (or) precipitation falling on impervious material—is mostly diverted to spreading basins or impounded at dams to enhance recharge. A lesser amount of recharge occurs as subsurface flow from adjacent basins or from fractures in the surrounding mountains (California Department of Water Resources, 2005a,b,c) or from return flow from other sources such as leakage from pipes.

Imported water for direct use and for artificial recharge in the FG study unit is delivered from several distant sources. Beginning in 1913, Los Angeles Water & Power began to deliver water from Owens Valley by means of the Los Angeles Aqueduct. In 1941, the Metropolitan Water District of Southern California began to deliver water from the Colorado River to Southern California. In 1972, the Department of Water Resources began to deliver water from the San Francisco Bay–Delta area by means of the West- and East-Branch State Water Project (California Department of Water Resources, 2005a,b,c).

The combined safe yield for all basins (the amount of water the basin can yield without producing unacceptable negative effects) in the FG study unit is approximately 314,000 acre-feet per year (acre-ft/yr). (California Department of Water Resources, 2003). Most groundwater extractions in the study areas are controlled. The court ordered adjudication of the San Fernando Valley, the Raymond, and the San Gabriel Valley groundwater basins is administered by separate Watermaster appointments (California Department of Water Resources, 2005a,b,c).

Methods

The *status assessment* provides a spatially-unbiased assessment of groundwater quality in the primary aquifers of the FG study unit. This section describes the methods used for: (1) defining groundwater quality, (2) assembling the datasets used for the status assessment, (3) determining which constituents warrant additional evaluation, and (4) calculating aquifer-scale proportions. Methods used for compilation of data regarding potential explanatory factors are described in appendix A.

The primary metric for defining groundwater quality is *relative-concentration*, which references concentrations

of constituents measured in groundwater to regulatory and non-regulatory benchmarks used to evaluate drinking-water quality. Some benchmarks are established for protection of human health and others are established for aesthetic properties, such as taste or odor. Constituents were selected for additional evaluation in the assessment on the basis of objective criteria by using these relative-concentrations. Groundwater-quality data collected by the U.S. Geological Survey for the GAMA Priority Basin Project (USGS–GAMA) and data compiled in the CDPH database are used in the *status assessment*. Two statistical approaches based on spatially unbiased equal-area grids are used to calculate aquifer-scale proportions of low, moderate, or high relative-concentrations (Belitz and others, 2010): (1) the "grid-based" approach uses one value per grid cell to represent groundwater quality, and (2) the "spatially weighted" approach uses many values per grid cell.

The CDPH database contains historical records from more than 25,000 wells, necessitating targeted retrievals to effectively access relevant water-quality data. For example, for the area representing the FG study unit, the historical CDPH database contains more than 1,400,000 records from 700 wells. The CDPH data were used in three ways in the *status assessment*: (1) to supplement the USGS data for the grid-based calculations of aquifer-scale proportions, (2) to select constituents for additional evaluation in the assessment, and (3) to provide the majority of the data used in the spatially weighted calculations of aquifer-scale proportions.

Relative-Concentrations and Water-Quality Benchmarks

Concentrations of constituents are presented as relative-concentrations in the *status assessment*:

$$\text{Relative-concentration} = \frac{\text{Sample concentration}}{\text{Benchmark concentration}}.$$

Relative-concentrations were used to provide context for the measured concentrations in the sample. Relative-concentrations less than 1 (<1.0) indicate a sample concentration less than the benchmark, and relative-concentrations greater than 1 (>1.0) indicate a sample concentration greater than the benchmark. The use of relative concentrations also permits comparison on a single scale of constituents present at a wide range of concentrations.

Toccalino and others (2004), Toccalino and Norman (2006), and Rowe and others (2007) previously used the ratio of measured sample concentration to the benchmark concentration [either maximum contaminant levels (MCLs) or Health-Based Screening Levels (HBSLs)] and defined this ratio as the Benchmark Quotient. Relative-concentrations used in this report are equivalent to the Benchmark Quotient reported by Toccalino and others (2004) for constituents with MCLs. However, HBSLs were not used in this report because HBSLs are not currently used as benchmarks by California drinking-water regulatory agencies. Relative-concentrations can only be computed for constituents with water-quality benchmarks; therefore, constituents without water-quality benchmarks are not included in the *status assessment*.

Regulatory and non-regulatory benchmarks apply to treated water that is served to the consumer, not to untreated groundwater. However, to provide some context for the results, concentrations of constituents measured in the untreated groundwater were compared to benchmarks established by the U.S. Environmental Protection Agency (USEPA) and CDPH (U.S. Environmental Protection Agency, 2006; California Department of Public Health, 2008a,b). The benchmarks used for each constituent were selected in the following order of priority:

1. Regulatory, health-based CDPH and USEPA maximum contaminant levels (MCL-CA and MCL-US), action levels (AL-US), and treatment technique levels (TT-US).

2. Non-regulatory CDPH and USEPA secondary maximum contaminant levels (SMCL-CA and SMCL-US). For constituents with both recommended and upper SMCL-CA levels, the values for the upper levels were used.

3. Non-regulatory, health-based CDPH notification levels (NL-CA), USEPA lifetime health-advisory levels (HAL-US) and USEPA risk-specific doses for 1:100,000 (RSD5-US).

For constituents with multiple types of benchmarks, this hierarchy may not result in selection of the benchmark with the lowest concentration. Additional information on the types of benchmarks and listings of the benchmarks for all constituents analyzed is provided by Land and Belitz (2008).

For ease of discussion, relative-concentrations of constituents were classified into low, moderate, and high categories:

Category	Relative-concentrations for organic and special-interest constituents	Relative-concentrations for inorganic constituents
High	> 1	> 1
Moderate	> 0.1 and ≤ 1	> 0.5 and ≤ 1
Low	≤ 0.1	≤ 0.5

For organic and special-interest constituents, a relative-concentration of 0.1 was used as a threshold to distinguish between low and moderate relative-concentrations for consistency with other studies and reporting requirements (U.S. Environmental Protection Agency, 1998; Toccalino and others, 2004). For inorganic constituents,

a relative-concentration of 0.5 was used as a threshold to distinguish between low and moderate relative-concentrations. A higher threshold value was used because in the FG study unit and elsewhere in California (Kulongoski and others, 2010; Landon and others, 2010; Kulongoski and Belitz, 2011), the naturally occurring inorganic constituents tend to be more prevalent than organic constituents in groundwater. Although more complex classifications could be devised based on the properties and sources of individual constituents, use of a single moderate/low threshold value for each of the two major groups of constituents provided a consistent objective criteria for distinguishing constituents at moderate rather than low concentrations.

Datasets for Status Assessment

U.S. Geological Survey Grid Wells

The primary data used for the grid-based calculations of aquifer-scale proportions of relative-concentrations were data from wells sampled by USGS for the GAMA Priority Basin Project (USGS-GAMA). Detailed descriptions of the methods used to identify wells for sampling are given in Land and Belitz (2008). Briefly, each study area was divided into 10-mi² (~25 km²) equal-area grid cells, and in each cell, one well was randomly selected for sampling to represent the cell (fig. 4) (Scott, 1990). In the San Fernando study area, to avoid cells with no wells, a 1.9-mi (3-km) buffer was drawn around each public-supply well, and these areas were aggregated into 10-mi² grid cells from which the wells were selected (Land and Belitz, 2008). Wells were selected to sample from the population of wells in statewide databases maintained by the CDPH and the USGS. The FG study unit contained a total of 40 grid cells, and the USGS sampled one well in each of 35 of those cells (USGS-grid wells). Of the 35 USGS-grid wells, 32 were listed in the CDPH database, 2 were industrial wells, and 1 was an irrigation well perforated at depths similar to the depths of CDPH wells in their respective cells. USGS-grid wells were named with an alphanumeric GAMA ID consisting of a prefix identifying the study area and a number indicating the order of sample collection (fig. B1A; table A1, A2). The following prefixes were used to identify the study area: SF, San Fernando study area, SG, San Gabriel study area.

Samples collected from USGS-grid wells were analyzed for 163 to 291 constituents (table 1). Water-quality indicators (field parameters), volatile organic compounds, pesticides, noble gases, and selected isotopes were analyzed in samples from all USGS wells. Samples from a subset of 24 wells were analyzed for major and minor ions, trace elements, nutrients, redox species, radiochemical constituents, carbon isotopes, *N*-nitrosodimethylamine (NDMA), perchlorate, and 1,2,3-trichloropropane (1,2,3-TCP). The collection, analysis, and quality-control data for the analytes listed in table 1 are described by Land and Belitz (2008).

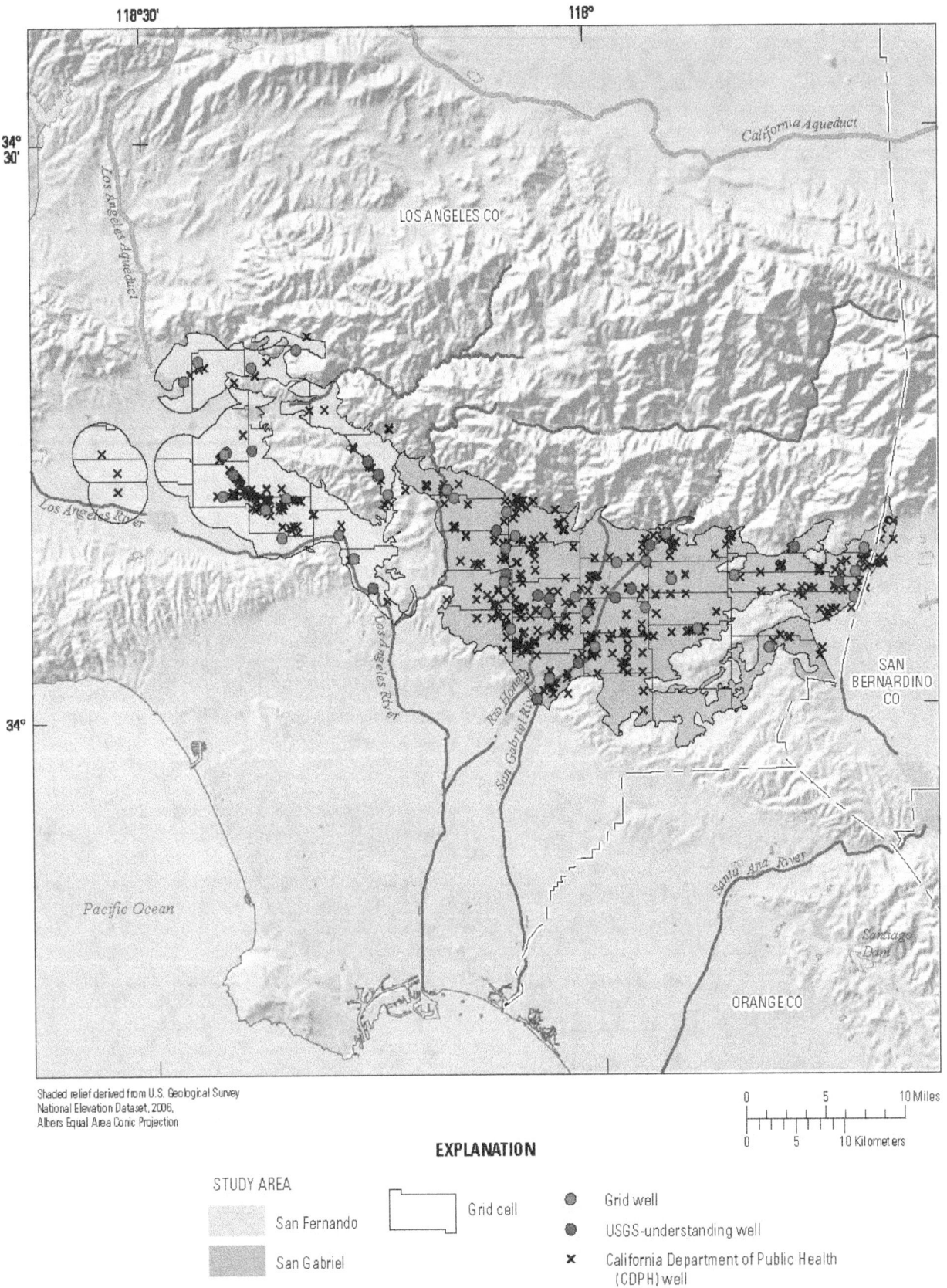

Figure 4.   Locations of study area grid cells, U.S. Geological Survey (USGS) grid and understanding wells, and California Department of Public Health (CDPH) wells, San Fernando–San Gabriel study unit, California GAMA Priority Basin Project, May–August 2005.

Table 1.   Number of wells sampled for the fast, intermediate, and slow sampling schedules, and number of constituents sampled in each constituent class, for the San Fernando–San Gabriel study unit, California GAMA Priority Basin Project, May–August 2005.

	Sampling schedule		
	Fast	Intermediate	Slow
Well summary	**Number of wells**		
Total number of wells	28	7	17
Number of grid wells sampled	24	3	8
Number of understanding wells sampled	4	4	9
Constituent class	**Number of constituents**		
Water-quality indicators (field parameters)			
Specific conductance, temperature, dissolved oxygen	3	3	3
Alkalinity, pH, turbidity, carbonate, bicarbonate			5
Organic compounds			
Volatile organic compounds (VOCs) and gasoline additives[1]	85	85	85
Pesticides and pesticide degradates	64	64	64
Polar pesticides and degradates			59
Constituents of special interest			
N-nitrosodimethylamine, perchlorate, 1,4-dioxane, and 1,2,3-Trichloropropane[2]		4	4
Inorganic constituents			
Major and minor ions, and trace elements		36	36
Nutrients plus dissolved organic carbon		6	6
Arsenic and iron species		4	4
Chromium species	2	2	2
Isotopes			
Stable isotopes of hydrogen and oxygen	2	2	2
Carbon isotopes			2
Radioactive constituents and dissolved gases			
Tritium[3]		1	1
Noble gases and tritium[4]	7	7	7
Radon and radium isotopes			3
Gross-alpha and beta radioactivity			4
Microbial constituents			
Total coliforms, colifage (somatic and F-specific), E. coli			4
Total	163	214	291

[1] Includes nine constituents classified as fumigants or fumigant synthesis byproducts.

[2] 1,2,3-TCP was analyzed as a constituent of special interest with a method reporting level of 0.005 µg/L (microgram per liter), and also on the USGS VOC schedule 2020, which has a laboratory reporting level of 0.12 µg/L.

[3] Analyzed at U.S. Geological Survey Tritium Laboratory, Menlo Park, California.

[4] Analyzed at Lawrence Livermore National Laboratory, Livermore, California.

California Department of Public Health Grid Wells

The two study areas were divided into 40 grid cells; of these, 5 cells did not have a USGS-grid well. Twenty-four cells had a USGS-grid well but no USGS data for major ions, trace elements, nutrients, and radiochemical constituents. The CDPH database was queried to provide these missing inorganic and radiochemical data. CDPH wells with data for the most recent 3 years available at the time of sampling (May 1 2002–April 30, 2005) were considered. If a well had more than one analysis for a constituent in the 3-year interval, then the most recent data were selected.

The procedures used to identify suitable data from CDPH wells are described in appendix B. Briefly, the first choice was to use CDPH data from the same well sampled by the USGS (USGS-grid well). In this case, "DG" was added to the well's GAMA ID to signify that it was a well sampled by the USGS that also had CDPH data (fig. B1B; table A2). If the DG well did not have all the needed data, then a second well in the cell was randomly selected from the subset of CDPH wells with data and a new identification with "DPH" and a new number was assigned to that well (fig. B1B; table A2). The combination of the USGS-grid wells and the DG- and DPH-CDPH-grid wells produced a grid-well network covering 35 of the 40 grid cells in the FG study unit (table A2). No accessible wells or data were available for the remaining 5 cells.

The CDPH database generally did not contain data for all missing inorganic constituents at every CDPH-grid well; therefore, the number of wells used for the grid-based assessment differed for various inorganic constituents (table 2). Although other organizations also collect water-quality data, the CDPH data is the only statewide database of groundwater-chemistry data available for comprehensive analysis.

CDPH data were not used to provide grid values for VOCs, pesticides, or perchlorate for the *status assessment* because a larger number of VOCs and pesticide compounds are analyzed for the USGS-GAMA Program than are available from the CDPH database. USGS-GAMA collected data for 85 VOCs plus 123 pesticides and pesticide degrades at each of the 52 wells sampled by the USGS in the FG study unit (table 1). In addition, method detection limits for USGS-GAMA analyses typically were one to two orders of magnitude less than the reporting levels for analyses compiled by the CDPH (table 3).

Additional Data Used for Spatially Weighted Calculations

The spatially weighted calculations of aquifer-scale proportions of relative-concentrations were made from data from the USGS-grid wells, from additional wells sampled by USGS-GAMA, and from all wells in the CDPH database with water-quality data during the 3-year interval May 1, 2002, to April 30, 2005. For wells with both USGS and CDPH data, only the USGS data were used.

Seventeen additional wells were sampled by the USGS to increase the sampling density in the FG study unit to better understand specific groundwater-quality issues. These "USGS-understanding" wells were numbered with prefixes modified from those used for the USGS-grid wells (for example, SFU-01to to SFU-06 and SGU-01to SGU-11) (fig. B1A; table A1, A2).

Selection of Constituents for Additional Evaluation

As many as 291 constituents were analyzed in samples from FG study unit wells; however, only a subset of these constituents are identified for additional evaluation in this report based on the following three criteria:

1. Constituents present at high or moderate relative concentrations in the CDPH database within the 3-year interval (May 1, 2002–April 30, 2005);

2. Constituents present at high or moderate relative-concentrations in the USGS-grid wells or USGS-understanding wells; or

3. Organic constituents with detection frequencies of greater than 10 percent in the USGS-grid well dataset for the study unit.

These criteria identified 22 organic constituents, 9 inorganic constituents, and 2 constituents of special interest for additional evaluation (table 4). An additional 42 organic constituents and 40 inorganic constituents were detected by USGS-GAMA, but were not selected for additional evaluation because either benchmarks were not established or detection was at low relative-concentrations (table 5). A complete list of the constituents investigated by USGS-GAMA in the FG study unit may be found in the FG Data Series Report (Land and Belitz, 2008).

Table 2. Inorganic constituents and associated benchmark information, and number of grid wells with U.S. Geological Survey-GAMA data and CDPH data, for each constituent, San Fernando–San Gabriel study unit, California GAMA Priority Basin Project.

[CDPH, California Department of Public Health; USEPA, U.S. Environmental Protection Agency; MCL-CA, California CDPH maximum contaminant level; MCL-US, USEPA maximum contaminant level; SMCL-CA, CDPH secondary maximum contaminant level; SMCL-US, USEPA secondary maximum contaminant level; NL-CA, CDPH notification level; AL-US, USEPA action level; HAL-US, USEPA lifetime health advisory level; USGS, U.S. Geological Survey]

Constituent	Benchmark type	Benchmark value	Number of grid wells with USGS-GAMA data	Number of grid wells with CDPH data
Nutrient, in milligrams per liter				
Ammonia, as nitrogen	HAL-US	30	11	4
Nitrate plus nitrite, as nitrogen	MCL-US	10	11	23
Nitrite, as nitrogen	MCL-US	1	11	21
Trace element, in micrograms per liter				
Aluminum	MCL-CA	1,000	11	23
Antimony	MCL-US	6	11	22
Arsenic	MCL-US	10	11	22
Barium	MCL-CA	1,000	11	22
Beryllium	MCL-US	4	11	21
Boron	NL-CA	1,000	11	19
Cadmium	MCL-US	5	11	22
Chromium	MCL-CA	50	11	23
Copper	AL-US	1,300	11	22
Iron	SMCL-CA	300	11	22
Lead	AL-US	15	11	21
Manganese	SMCL-CA	50	11	22
Mercury	MCL-US	2	11	22
Molybdenum	HAL-US	40	11	0
Nickel	MCL-CA	100	11	22
Selenium	MCL-US	50	11	22
Silver	SMCL-CA	100	11	22
Strontium	HAL-US	4,000	16	0
Thallium	MCL-US	2	11	21
Uranium	MCL-US	30	11	9
Vanadium	NL-CA	50	11	9
Zinc	SMCL-US	5,000	11	22
Major ion and total dissolved solids, in milligrams per liter				
Chloride	SMCL-CA	500	11	22
Fluoride	MCL-CA	2	11	21
Sulfate	SMCL-CA	500	11	21
Total dissolved solids (TDS)	SMCL-CA	1,000	11	22
Radioactive constituents, in picocuries per liter				
Gross-alpha radioactivity, 30-day count	MCL-US	15	8	20
Gross-beta radioactivity, 30-day count	MCL-CA	50	8	5
Radium-226	MCL-US	5	8	20
Radium-228	MCL-US	5	8	20
Radon-222	MCL-US	4,000	7	0

Table 3. Comparison of the number of compounds and median laboratory reporting levels or method detection limits by type of constituent for data reported in the California Department of Public Health (CDPH) database and for data collected by the U.S. Geological Survey (USGS) for the San Fernando–San Gabriel study unit, California GAMA Priority Basin Project, May–August 2005.

[MDL, method detection limit; LRL, laboratory reporting level; mg/L, milligrams per liter; µg/L, micrograms per liter; pCi/L, picocuries per liter; SSMDC, sample-specific minimum detectable concentration; na, not available]

Constituent type	CDPH		GAMA	
	Number of compounds	Median MDL	Number of compounds	Median LRL
Organic constituents				
Volatile organic compounds plus gasoline additives (including fumigants)	65	0.5	88	0.06
Pesticides plus degradates	21	3	112	0.018
Inorganic constituents				
Nutrients, major and minor ions	14	na	11	0.04
Trace elements	19	6	25	0.12
Radioactive constituents (SSMDC)	5	1	[1]7	0.42
Constituents of special interest				
Perchlorate	1	4	1	0.5
1,2,3-Trichloropropane (TCP)	1	0.5	1	0.005
1,4-Dioxane	1	na	1	2.0
N-Nitrosodimethylamine (NDMA)	1	0.002	1	0.002

[1] Excludes tritium.

Table 4. Aquifer-scale proportions from grid-based and spatially weighted approaches for constituents with moderate or high relative-concentrations during May 1, 2002–April 30, 2005, from the California Department of Public Health (CDPH) database, or with moderate or high relative-concentrations in samples collected from USGS-grid wells (May–August 2005), San Fernando–San Gabriel study unit, California GAMA Priority Basin Project.

[CDPH, California Department of Public Health; USEPA, U.S. Environmental Protection Agency. Aquifer-scale proportions are areal. Benchmark value units: trace elements and minor elements, solvents, other VOCs, trihalomethanes, herbicides, fumigants, and constituents of special interest, micrograms per liter (µg/L); radioactive constituents, picocuries per liter (pCi/L); nutrients, inorganic constituents with SCML thresholds, milligrams per liter (mg/L). Grid-based aquifer proportions for organic constituents are based on samples collected by the U.S. Geological Survey from 35 grid wells during May 24–July 20, 2005. Spatially weighted aquifer proportions are based on CDPH data during May 1, 2002–April 30, 2005, in combination with grid well and understanding well data. High, concentrations greater than benchmark; moderate, concentrations less than benchmark and greater than or equal to 0.1 (for organic constituents) or 0.5 (for inorganic constituents) of benchmark; low, concentrations less than 0.1 (for organic constituents) or 0.5 (for inorganic constituents) of benchmark. MCL-CA, CDPH maximum contaminant level; MCL-US, USEPA maximum contaminant level; NL-CA, CDPH notification level; AL-US, USEPA action level; SMCL-CA, CDPH secondary maximum contaminant level]

Constituent	Benchmark Type	Benchmark Value	Percent detection frequency[1] Number of wells	Moderate values (percent)	High values (percent)	Spatially weighted aquifer-scale proportion of moderate and high relative-concentrations[1] Number of cells	Moderate values (percent)	High values (percent)	Grid-based aquifer proportion of moderate and high relative-concentrations Number of cells	Moderate values (percent)	High values (percent)	90-percent confidence interval for grid-based high proportion[2] Lower limit (percent)	Upper limit (percent)
Solvents													
Carbon tetrachloride	MCL-CA	0.5	403	1.7	6.4	35	2.3	6.5	35	11.4	5.7	1.9	15.9
1,1-Dichloroethane	MCL-CA	5	404	4.0	0.5	35	3.5	0.5	35	0.0	0.0	0.0	4.5
1,2-Dichloroethane	MCL-CA	0.5	404	1.0	2.0	35	1.3	1.9	35	5.7	0.0	0.0	4.5
cis-1,2-Dichloroethene	MCL-CA	6	404	7.2	1.0	35	6.6	1.0	35	2.9	0.0	0.0	4.5
Perchloroethene (PCE)	MCL-US	5	406	23.6	13.1	35	19.0	11.2	35	17.1	8.6	3.5	19.6
Trichloroethene (TCE)	MCL-US	5	405	21.0	14.6	35	15.0	14.8	35	14.3	8.6	3.5	19.6
Trihalomethanes													
Bromodichloromethane (THM)	MCL-US	80	404	1.0	0.0	35	0.7	0.0	35	0.0	0.0	0.0	4.5
Chloroform (THM)[3]	MCL-US	80	404	2.0	0.0	35	2.1	0.0	35	0.0	0.0	0.0	4.5
Dibromochloromethane (THM)	MCL-US	80	404	0.5	0.0	35	0.5	0.0	35	0.0	0.0	0.0	4.5
Total trihalomethanes[3]	MCL-US	100	404	3.5	0.0	35	3.3	0.0	35	0.0	0.0	0.0	4.5
Other VOCs													
CFC-11	MCL-CA	150	404	0.2	0.0	35	0.2	0.0	35	0.0	0.0	0.0	4.5
Total chlorofluorocarbons (CFCs)	MCL-CA	150	404	0.2	0.0	35	0.2	0.0	35	0.0	0.0	0.0	4.5
1,1-Dichloroethene	MCL-CA	6	405	4.2	2.2	35	5.2	2.9	35	5.7	0.0	0.0	4.5
Methyl tert-butyl ether (MTBE)	MCL-CA	13	405	0.2	0.0	35	0.2	0.0	35	0.0	0.0	0.0	4.5
Constituents of special interest													
Perchlorate	MCL-CA	6	382	10.5	8.6	34	12.0	11.2	11	36.4	0.0	0.0	13.0
N-Nitrosodimethylamine (NDMA)	NL-CA	0.010	72	9.7	8.3	20	7.9	5.2	11	9.1	0.0	0.0	13.0
Herbicides													
Atrazine	MCL-CA	1	375	0.5	0.0	35	3.1	0.0	35	5.7	0.0	0.0	4.5

Table 4. Aquifer-scale proportions from grid-based and spatially weighted approaches for constituents with moderate or high relative-concentrations during May 1, 2002–April 30, 2005, from the California Department of Public Health (CDPH) database, or with moderate or high relative-concentrations in samples collected from USGS-grid wells (May–August 2005), San Fernando–San Gabriel study unit, California GAMA Priority Basin Project.—Continued

[CDPH, California Department of Public Health; USEPA, U.S. Environmental Protection Agency. Aquifer-scale proportions are areal. Benchmark value units: trace elements and minor elements, solvents, other VOCs, trihalomethanes, herbicides, fumigants, and constituents of special interest, micrograms per liter (µg/L); radioactive constituents, picocuries per liter (pCi/L); nutrients, inorganic constituents with SCML thresholds, milligrams per liter (mg/L). Grid-based aquifer-scale proportions for organic constituents are based on samples collected by the U.S. Geological Survey from 35 grid wells during May 24–July 20, 2005. Spatially weighted aquifer proportions are based on CDPH data during May 1, 2002–April 30, 2005, in combination with grid well and understanding well data. High, concentrations greater than benchmark; moderate, concentrations less than benchmark and greater than or equal to 0.1 (for organic constituents) or 0.5 (for inorganic constituents) of benchmark; low, concentrations less than 0.1 (for organic constituents) or 0.5 (for inorganic constituents) of benchmark. MCL-CA, CDPH maximum contaminant level; NL-CA, CDPH notification level; AL-US, USEPA action level; SMCL-CA, CDPH secondary maximum contaminant level]

Constituent	Benchmark Type	Benchmark Value	Percent detection frequency[1] Number of wells	Percent detection frequency[1] Moderate values (percent)	Percent detection frequency[1] High values (percent)	Spatially weighted aquifer-scale proportion of moderate and high relative-concentrations[1] Number of cells	Spatially weighted Moderate values (percent)	Spatially weighted High values (percent)	Grid-based aquifer proportion of moderate and high relative-concentrations Number of cells	Grid-based Moderate values (percent)	Grid-based High values (percent)	90-percent confidence interval for grid-based high proportion[2] Lower limit (percent)	90-percent confidence interval Upper limit (percent)
Fumigants													
Dibromochloropropane (DBCP)[3]	MCL-US	0.2	404	0.0	0.0	35	0.0	0.0	35	0.0	0.0	0.0	4.5
Inorganic trace and minor elements													
Aluminum[3]	MCL-CA	1,000	370	0.0	0.0	34	0.0	0.0	34	0.0	0.0	0.0	4.6
Chromium	MCL-CA	50	373	1.1	0.8	34	1.4	1.3	34	2.9	0.0	0.0	4.6
Fluoride	MCL-CA	2	367	4.6	0.8	33	4.3	0.7	32	3.1	0.0	0.0	4.9
Lead[3]	AL-US	15	371	0.0	0.0	34	0.0	0.0	32	0.0	0.0	0.0	4.9
Uranium	MCL-US	30	125	16.0	0.0	31	13.7	0.0	20	10.0	0.0	0.0	7.6
Vanadium[3]	NL-CA	50	69	0.0	0.0	31	0.0	0.0	20	0.0	0.0	0.0	7.6
Inorganic constituents with SMCL benchmarks													
Iron	SMCL-CA	300	371	4.0	4.6	33	4.3	4.6	33	0.0	3.0	0.6	9.4
Total dissolved solids (TDS)	SMCL-CA	1,000	364	22.8	0.5	33	23.3	0.2	33	18.2	0.0	0.0	4.7
Sulfate	SMCL-CA	500	357	4.5	0.0	33	1.8	0.0	32	3.1	0.0	0.0	4.9
Nutrients													
Nitrate, as nitrogen	MCL-US	10	402	29.1	9.0	34	33.5	12.3	34	26.5	8.8	3.6	17.1
Radioactive constituents													
Gross-alpha radioactivity, 30-day count	MCL-US	15	356	11.5	0.6	33	1.0	0.3	28	0.0	0.0	0.0	5.5
Combined radium-226 plus radium-228[3]	MCL-US	5	333	0.0	0.0	33	0.0	0.0	28	0.0	0.0	0.0	5.5

[1] Based on the most recent data for each CDPH well during the period May 1, 2002–April 30, 2005, combined with GAMA grid and understanding well data.

[2] Based on the Jeffreys interval for the binomial distribution (Brown and others, 2001).

[3] The high value was reported in the CDPH database between May 1, 2002, and April 30, 2005, but this high value was not the most recently reported value used for calculating aquifer-scale proportion.

Table 5. Number of constituents analyzed and detected by the U.S. Geological Survey with associated benchmarks in each constituent class, San Fernando–San Gabriel study unit, California GAMA Priority Basin Project, May–August 2005.

[Health-based benchmarks include U.S. Environmental Protection Agency (USEPA) and California Department of Public Health (CDPH) maximum contaminant levels (MCL), USEPA lifetime health advisory levels (HAL) and risk-specific dose level at 10^{-5} lifetime cancer risk, and CDPH notification level (NL); RSD5, USEPA risk specific does at 10^{-5}; AL, USEPA action level; SMCL, USEPA or CDPH secondary maximum contaminant level. VOC, volatile organic compound]

Organic constituents class

Benchmark type	Sum of organic and special-interest constituents		VOC and gasoline additives (excluding fumigants)		Fumigants		Pesticides and degradates		Polar pesticides and degradates		Special interest	
	Analyzed	Detected	Analyzed	Detected	Analyzed	Detected	Analyzed	Detected	Analyzed	Detected	Analyzed	Detected
MCL	47	31	29	25	4	2	3	2	10	1	1	1
HAL	33	11	6	2	1	0	14	4	11	4	1	1
NL	17	5	15	4	0	0	0	0	0	0	2	1
RSD5	8	1	2	0	2	0	3	1	1	0	0	0
AL	0	0	0	0	0	0	0	0	0	0	0	0
SMCL	0	0	0	0	0	0	0	0	0	0	0	0
None	107	18	24	2	2	0	44	8	37	8	0	0
Total:	212	66	76	33	9	2	64	15	59	13	4	3

Inorganic constituents class

Benchmark type	Sum of inorganic constituents		Major and minor ions		Nutrients plus dissolved organic carbon		Trace elements		Radioactive constituents	
	Analyzed	Detected	Analyzed	Detected	Analyzed	Detected	Analyzed	Detected	Analyzed	Detected
MCL	24	24	1	1	2	2	12	12	8	8
HAL	5	4	0	0	1	0	3	3	0	0
NL	4	3	0	0	0	0	2	2	0	0
RSD5	0	0	0	0	0	0	0	0	0	0
AL	2	2	0	0	0	0	2	2	0	0
SMCL	6	6	3	3	0	0	3	3	0	0
None	13	13	7	7	3	3	3	3	0	0
Total:	50	49	11	11	6	5	25	25	8	8

Organic and inorganic constituents combined

	Analyzed	Detected
Total:	262	115

The CDPH database also was used to identify constituents with high relative-concentrations historically, but not currently. The historical period was defined as from the earliest record maintained in the CDPH database to April 30, 2002 (February 2, 1976–April 30, 2002).

Constituent concentrations may be historically high, but not currently high, because of improvement of groundwater quality with time or abandonment of wells with high concentrations. Historically high concentrations of constituents that do not otherwise meet the criteria for additional evaluation are not considered representative of potential groundwater-quality concerns in the study unit from 2002 to 2005. For the FG study unit, 24 constituents were identified at historically high concentrations (table 6).

Table 6. Constituents in the California Department of Public Health (CDPH) database at high concentrations from February 2, 1976– April 30, 2002, San Fernando–San Gabriel study unit, California GAMA Priority Basin Project.

[Benchmarks and benchmark values as of June 1, 2005. Benchmark type: MCL-CA, California Department of Public Health maximum contaminant level; MCL-US, U.S. Environmental Protection Agency maximum contaminant level; SMCL-CA, California Department of Public Health secondary maximum contaminant level; HAL-US, U.S. Environmental Protection Agency health advisory level; NL-CA, California Department of Public Health notification level. Relative concentration equals measured concentration divided by benchmark value; relative concentration greater than 1 is defined as high; µg/L, microgram per liter]

Constituent	Benchmark		Number of wells with an analysis	Number of wells with historically high concentrations	Date of most recent high value
	Type	Value (µg/L)			
Organic constituents					
1,1,1-trichloroethane	MCL-US	200	560	1	12-07-93
1,1,2-trichloroethane (1,1,2-TCA)	MCL-US	5	560	7	01-21-02
1,2,3-trichloropropane (1,2,3-TCP)	HAL-US	40	528	1	07-16-01
1,4-dioxane	NL-CA	3	19	4	09-19-01
Benzene	MCL-CA	1	560	6	03-05-96
bis(2-ethylhexyl) phthalate (DEHP)	MCL-CA	4	433	1	03-11-93
Dichloromethane (methylene chloride)	MCL-US	5	562	12	06-05-01
1,2-dibromoethane (EDB)	MCL-US	0.05	492	1	10-22-96
Heptachlor	MCL-CA	0.01	421	2	09-21-94
Lindane	MCL-US	0.2	450	1	04-20-89
trans-1,2-dichloroethene	MCL-CA	10	560	3	02-03-00
Vinyl chloride	MCL-CA	0.5	559	3	04-02-02
Inorganic constituents					
Antimony	MCL-US	6	469	1	06-06-95
Arsenic	MCL-US	10	517	13	03-13-02
Cadmium	MCL-US	5	517	1	11-09-89
Chromium (hexavalent)	MCL-CA	50	356	3	03-06-02
Manganese	SMCL-CA	50	523	23	02-14-02
Mercury	MCL-US	2	516	6	02-20-01
Molybdenum	HAL-US	40	85	3	02-25-00
Nickel	MCL-CA	100	471	1	07-26-00
Nitrite (as nitrogen)	MCL-US	1	463	1	09-28-99
Gross beta	MCL-CA	50	226	1	12-28-89
Radium-226	MCL-US	5	142	1	09-11-89
Radium-228	MCL-US	5	66	2	01-29-97

Calculation of Aquifer-Scale Proportions

The *status assessment* is intended to characterize the quality of groundwater resources in the primary aquifers of the FG study unit. The primary aquifers are defined by the perforated depth intervals of the wells listed in the CDPH database. The use of the term "primary aquifers" does not imply that there exists a discrete aquifer unit. In most groundwater basins, municipal and community supply wells generally are perforated at greater depths than domestic wells. Thus, because domestic wells are not listed in the CDPH database, the primary aquifers generally correspond to the portion of the aquifer system tapped by municipal and community supply wells. A majority of the wells used in the status assessment are listed in the CDPH database, and are therefore classified as municipal and community drinking-water supply wells. However, to the extent that domestic wells are perforated over the same depth intervals as the CDPH wells, the assessments presented in this report also may be applicable to the portions of the aquifer systems used for domestic drinking-water supplies.

Two statistical approaches, grid-based and spatially weighted (Belitz and others, 2010), were selected to evaluate the proportions of the primary aquifers in the FG study unit with high, moderate, and low relative-concentrations of constituents relative to benchmarks. For ease of discussion, these proportions are referred to as "high," "moderate," and "low" aquifer-scale proportions. Calculations of aquifer-scale proportions were made for individual constituents meeting the criteria for additional evaluation in the *status assessment*, as well as for classes of constituents. Classes of constituents with health-based benchmarks included: trace elements, radioactive constituents, nutrients, major and minor ions, solvents, trihalomethanes, other VOCs, herbicides, and special-interest constituents. Aquifer-scale proportions also were calculated for the following constituents with aesthetic (SMCL) benchmarks: total dissolved solids (TDS), chloride, manganese, iron, sulfate, and zinc.

The grid-based calculation uses the grid-well dataset assembled from the USGS-grid and CDPH-grid wells (Belitz and others, 2010). For each constituent, the high aquifer-scale proportion was calculated by dividing the number of cells represented by a high value for that constituent by the total number of grid cells with data for that constituent. The moderate and low aquifer-scale proportions were calculated similarly. Confidence intervals for the high aquifer-scale proportions were computed using the Jeffreys interval for the binomial distribution (Brown and others, 2001). Additional justification for this method was provided in Belitz and others, 2010. The grid-based estimate is spatially unbiased. However, the grid-based approach may not identify constituents that are present at high concentrations in small proportions of the primary aquifers. For calculation of high aquifer-scale proportion for a class of constituents, cells were considered high if values for any of the constituents in that class were high. Cells were considered moderate if values for any of the constituents were moderate and if no values were high.

The spatially weighted calculation uses the dataset assembled from all CDPH and USGS GAMA wells. For each constituent, the high aquifer-scale proportion was calculated by computing the proportion of wells with high values in each cell and then averaging the proportions for all cells (Belitz and others, 2010). The moderate aquifer-scale proportion was calculated similarly. Confidence intervals for spatially weighted detection frequencies of high concentrations are not described in this report. For calculation of high aquifer-scale proportion for a class of constituents, values for wells were considered high if the values for any of the constituents in that class were high. Values for wells were considered moderate if the values for any of the constituents were moderate and if no values for wells were high.

In addition, for each constituent, the raw detection frequencies of high and moderate values for individual constituents were calculated using the same dataset as used for the spatially weighted calculations. However, raw detection frequencies are not spatially unbiased because the wells in the CDPH database are not uniformly distributed throughout the FG study unit (fig. 4). For example, if a constituent were present at high concentrations in a small region of the aquifer with a high density of wells, the raw detection frequency of high values would be greater than the high aquifer-scale proportion. Raw detection frequencies are provided for reference but were not used to assess aquifer-scale proportions (see appendix C for details of statistical approaches).

The grid-based high aquifer-scale proportions were used to represent proportions in the primary aquifers unless the spatially weighted proportions were significantly different from the grid-based values. Significantly different results were defined as follows:

- If the grid-based high aquifer-scale proportion was zero and the spatially weighted proportion was non-zero, then the spatially weighted result was used. This situation can happen when the concentration of a constituent is high in a small fraction of the primary aquifers.

- If the grid-based high aquifer-scale proportion was non-zero and the spatially weighted proportion was outside the 90-percent confidence interval (based on the Jeffreys interval for the binomial distribution), then the spatially weighted proportion was used.

The grid-based moderate and low proportions were used in most cases because the reporting levels for many organic constituents and some inorganic constituents in the CDPH database were higher than the threshold between moderate and low categories. However, if the grid-based moderate proportion was zero and the spatially weighted proportion non-zero, then the spatially weighted value was used as a minimum estimate for the moderate proportion. Given the prevalence of perchloroethene (PCE), carbon tetrachloride, and trichloroethene (TCE) at high relative-concentrations in CDPH wells in the SG study area, and the absence of these constituents at high relative concentrations in the SG grid wells, the spatially weighted approach was used for these constituents.

Potential Explanatory Factors

Land Use

Land use in the FG study unit is 83 percent urban, 16 percent natural, and 1 percent agricultural, based on classifications from USGS National Land Cover Data (Nakagaki and others, 2007) (figs. 5, 6; appendix A). Urban land use is predominant and uniformly distributed across the SF and the SG study areas. The urban landscape consists primarily of low-intensity residential areas, followed by commercial, industrial, transportation, and high-intensity residential areas. Natural lands are mostly shrub lands, grasslands, or forests. Figure 5A shows the percentage of land use calculated for the entire study area and also for the area in a 500-meter (m) (1,640-ft) radius around each grid well. Figure 5B shows the land-use percentages at the well locations in the FG study unit. Figure 6 shows the land-use classification map based on satellite imagery (appendix A; table A1) and the locations of FG grid and understanding wells. A 500-m buffer surrounding the well has been shown to be effective at correlating urban land use with VOC occurrence for the purposes of statistical characterization (Johnson and Belitz, 2009).

Well Depth and Depth to Top-of-Perforation

Well construction information was available for 35 grid wells sampled in the FG study unit (table A3). Depths of grid wells ranged from 184 to 1,290 ft below land surface (BLS); the median was 544 ft BLS (fig. 7; table A3). Depths to the top-of-perforation ranged from 50 to 1,013 ft BLS, with a median of 230 ft BLS. The perforation length was as much as 650 ft, with a median of 300 ft. Well construction information also was available for 17 understanding wells. The understanding well depths, perforation lengths, and depth to top perforation (fig. 7) were comparable to those of the grid wells.

Groundwater Age

Groundwater samples were assigned age classifications on the basis of the tritium, carbon-14, and helium-4 content of the samples (see "Groundwater-Age Classification" in appendix A). Age classifications were assigned to 24 USGS-grid and understanding well samples: 13 were classified as modern, 6 were mixed, and 5 were pre-modern age (table A4).

Groundwater ages generally increased as depths to top-of-well perforations increased (fig. 8A). The depths to the top of perforations were generally shallower in wells with water classified as modern age and mixed age compared to those with water classified as pre-modern age. Water classified as modern was generally from shallower depths than for water classified as pre-modern (fig. 8B). Water in five of the nine wells perforated entirely within the upper 400 ft of the aquifer was modern age, whereas water in two wells (out of four) with perforations equal to or greater than 400 ft below land surface was pre-modern (fig. 8C).

Geochemical Condition

An abridged classification of oxidation-reduction (redox) conditions adapted from the framework presented by McMahon and Chapelle (2008) for 45 FG study unit wells sampled by the USGS-GAMA Priority Basin Project, and data from 7 wells reported in the CDPH database, is given in appendix A (table A5). The classification "indeterminate" was added to the framework for groundwater samples that did not have sufficient data available to be classified as oxic, anoxic/suboxic, or mixed (anoxic/oxic) (Jurgens and others, 2009). Groundwater was oxic in 96 percent of the wells, mixed (anoxic/oxic) in 2 percent of the wells, and indeterminate in 1 percent of the wells.

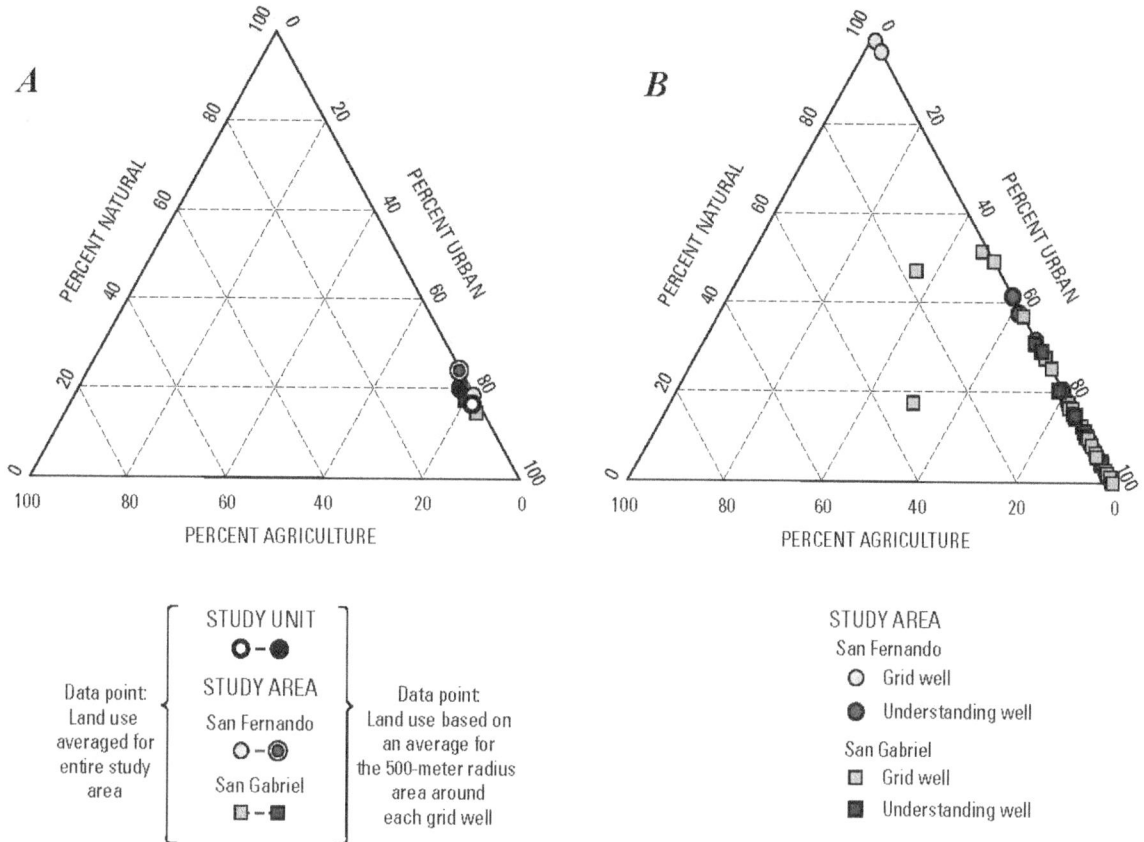

Figure 5.   Percentage of urban, agricultural, and natural land use in (*A*) the study unit and study areas based on land use in the study unit and the 500-meter-radius area surrounding each well, and (*B*) land-use classification for individual GAMA wells in the San Fernando–San Gabriel study unit, California GAMA Priority Basin Project.

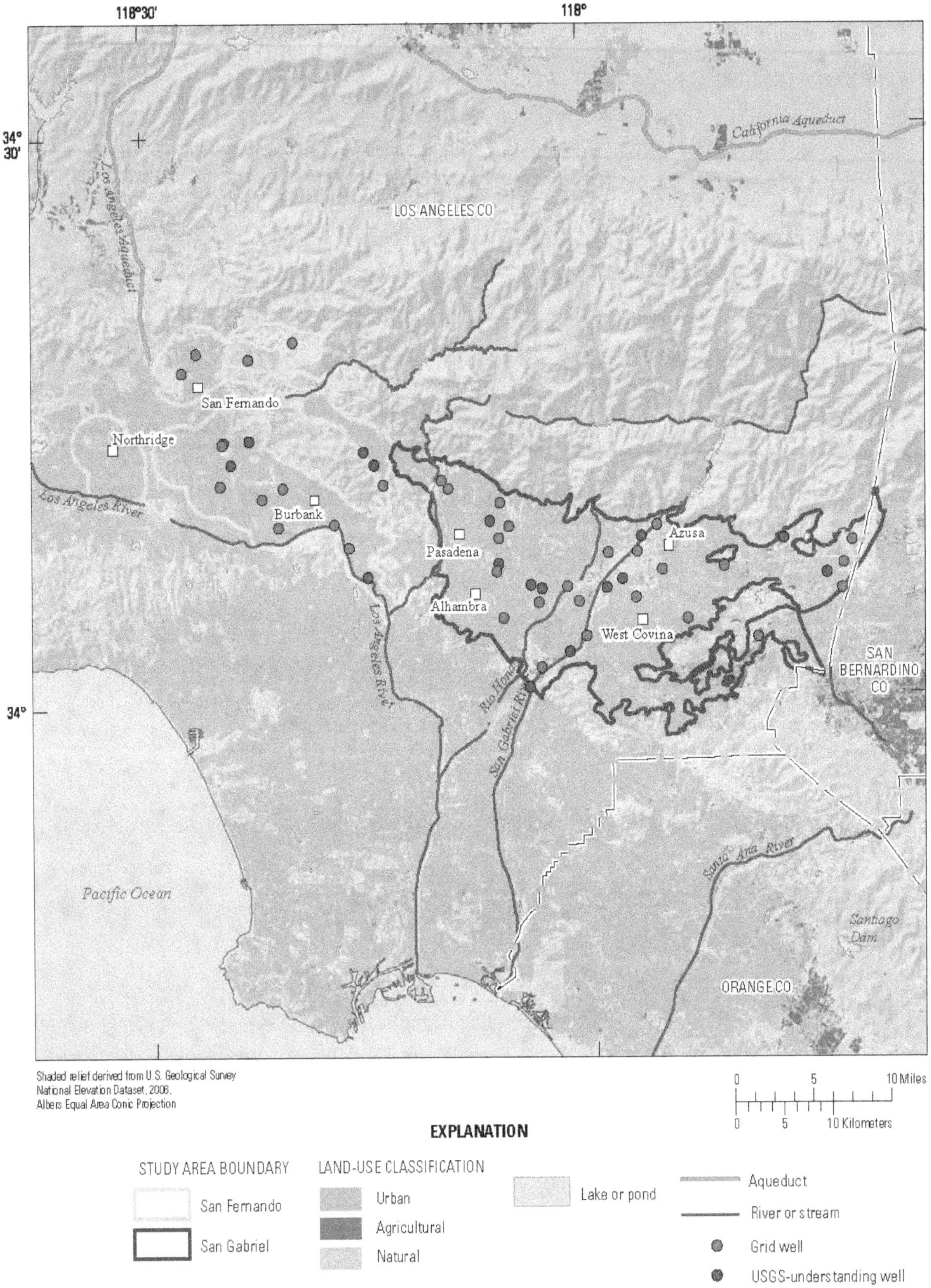
118°30'
118°
34° 30'
34°
California Aqueduct
Los Angeles Aqueduct
LOS ANGELES CO
San Fernando
Northridge
Los Angeles River
Burbank
Pasadena
Azusa
Alhambra
West Covina
Los Angeles River
Rio Hondo
San Gabriel River
SAN BERNARDINO CO
Santa Ana River
Pacific Ocean
Santiago Dam
ORANGE CO
Shaded relief derived from U.S. Geological Survey National Elevation Dataset, 2006, Albers Equal Area Conic Projection
0 5 10 Miles
0 5 10 Kilometers
EXPLANATION
STUDY AREA BOUNDARY
San Fernando
San Gabriel
LAND-USE CLASSIFICATION
Urban
Agricultural
Natural
Lake or pond
Aqueduct
River or stream
Grid well
USGS-understanding well

Figure 6. Land use and the locations of USGS-grid and understanding wells in the San Fernando–San Gabriel study unit, California GAMA Priority Basin Project.

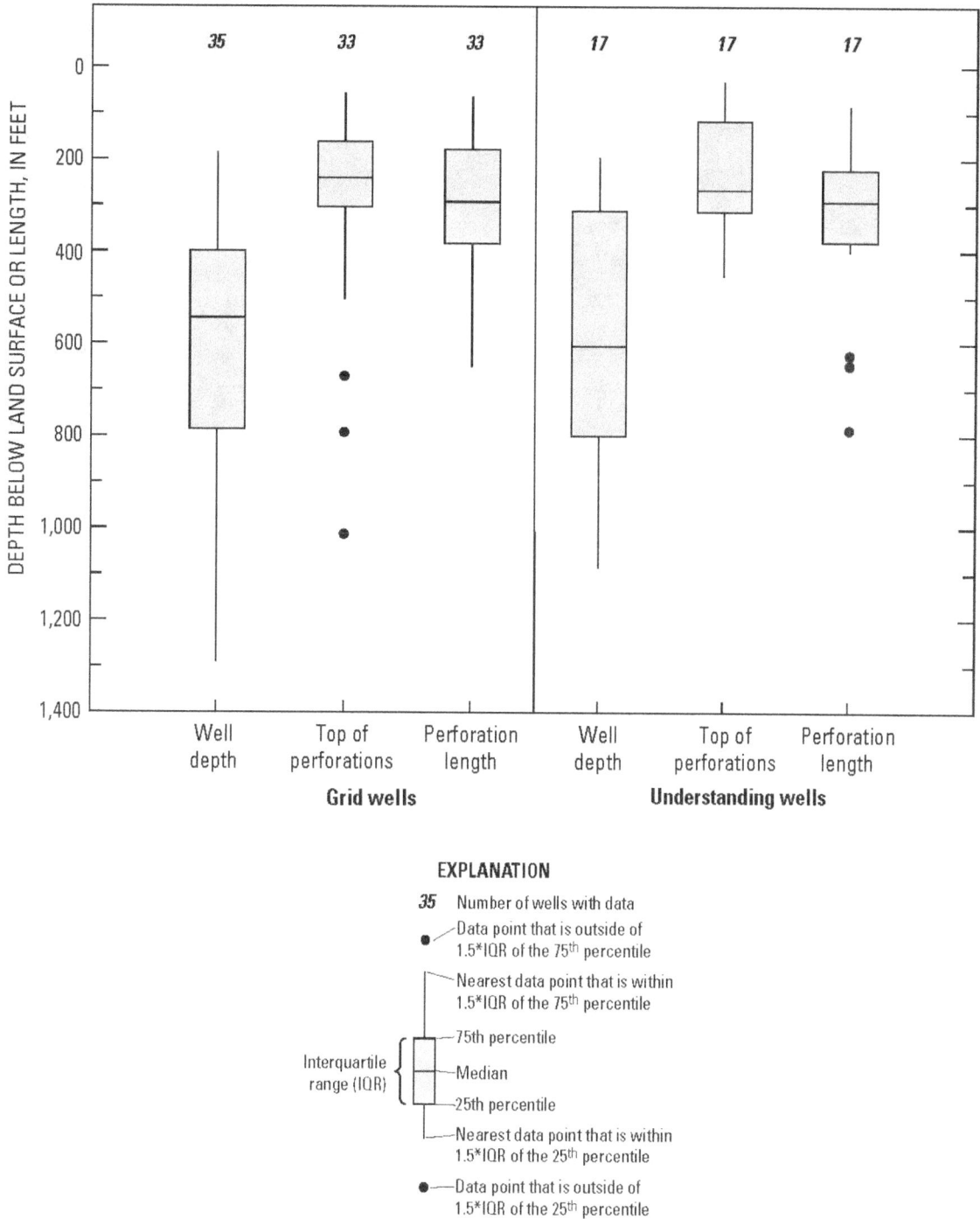

EXPLANATION

35   Number of wells with data

●   Data point that is outside of 1.5*IQR of the 75th percentile

│   Nearest data point that is within 1.5*IQR of the 75th percentile

75th percentile

Interquartile range (IQR)   Median

25th percentile

│   Nearest data point that is within 1.5*IQR of the 25th percentile

●   Data point that is outside of 1.5*IQR of the 25th percentile

Figure 7.   Well depths, depths to top-of-perforation, and perforation lengths for USGS-grid and understanding wells, San Fernando–San Gabriel study unit, California GAMA Priority Basin Project.

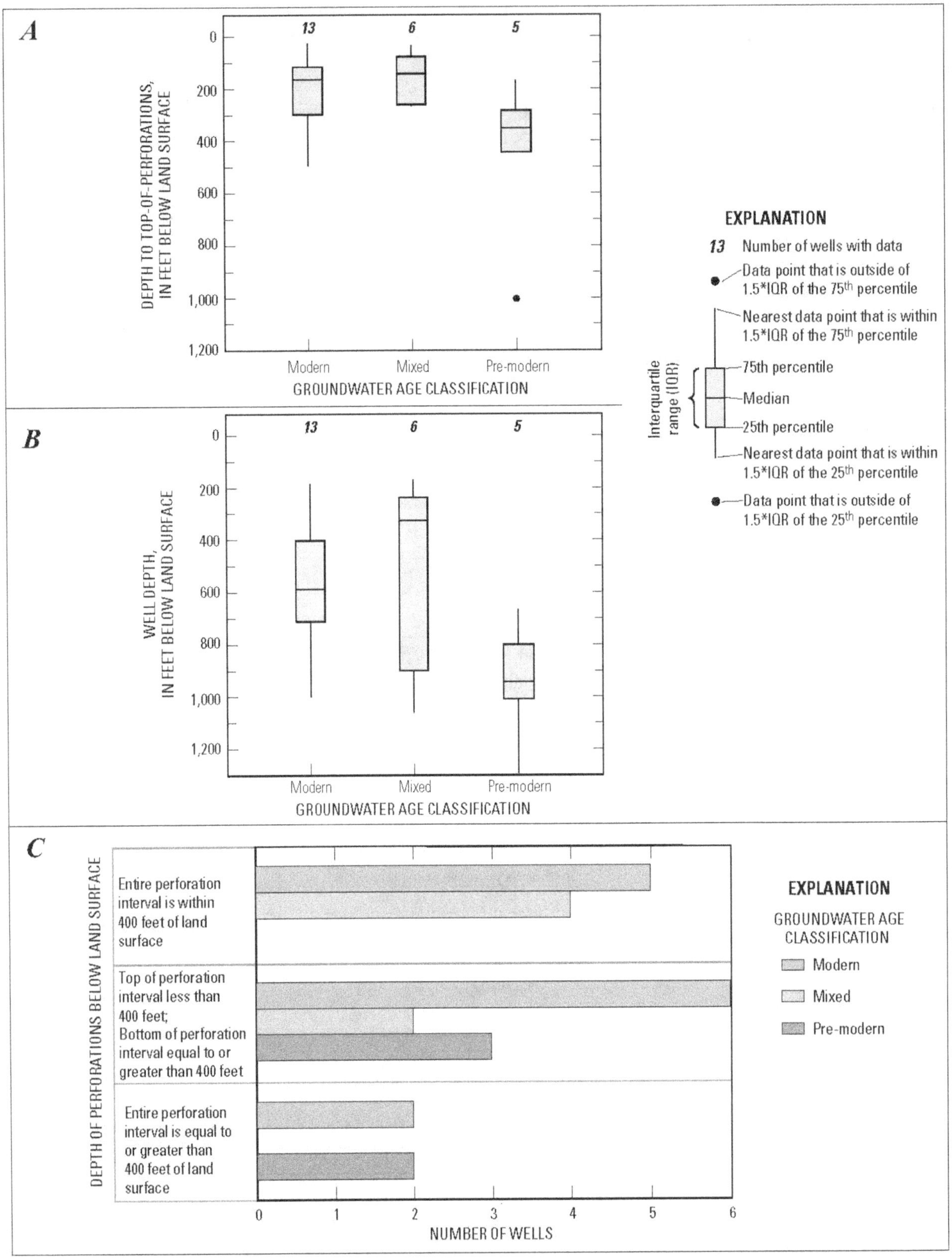

Figure 8.   Relation of groundwater age classification to (*A*) depth to top-of-perforations, (*B*) well depth, and (*C*) age classification, in relation to the depth of well perforations, San Fernando–San Gabriel study unit, California GAMA Priority Basin Project.

Status of Water Quality

The *status assessment* was designed to identify the constituents or classes of constituents most likely to be of water-quality concern because of their high relative-concentrations or their prevalence. USGS sample analyses, plus additional data from the CDPH database, were included in the assessment of groundwater quality for the FG study unit. The spatially distributed, randomized approach to grid-well selection and data analysis yields a view of groundwater quality in which all areas of the primary aquifers are weighted equally; regions with a high density of groundwater use or with high density of potential contaminants were not preferentially sampled (Belitz and others, 2010).

The following discussion of the *status assessment* results is divided into inorganic and organic constituents. The assessment begins with a survey of how many constituents were detected at any concentration compared to the number analyzed, and a graphical summary of the relative-concentrations of constituents detected in the grid wells. Results are presented for the subset of constituents that met criteria for selection for additional evaluation based on concentration, or for organic constituents, prevalence.

The aquifer-scale proportions calculated by using the spatially weighted approach were within the 90-percent confidence intervals for their respective grid-based aquifer high proportions for 30 of the 30 constituents listed in table 4, providing evidence that the grid-based and spatially weighted approaches yield statistically equivalent results.

Inorganic Constituents

Inorganic constituents generally occur naturally in groundwater, although their concentrations may be influenced by human factors as well as natural factors. Of the 50 inorganic constituents analyzed by the USGS-GAMA, 46 were detected in the FG study unit. Of the 50 inorganic constituents analyzed, 31 had regulatory or non-regulatory health-based benchmarks, 6 had non-regulatory aesthetic/ technical-based benchmarks, and 13 had no established benchmarks (table 5). The inorganic constituents detected at high relative-concentrations in one or more of the 35 grid wells were iron and nitrate (table 4). Four additional inorganic constituents were present at high-relative concentrations in the FG study unit but not in grid wells: chromium, fluoride, total dissolved solids, and gross-alpha radioactivity (table 4). The maximum relative-concentration (sample concentration divided by the benchmark concentration) for each constituent in grid wells is shown in figure 9.

Eight inorganic constituents—the trace elements chromium, uranium, iron; the minor element fluoride; the major ion sulfate; TDS; gross-alpha radioactivity (30-day

count); and the nutrient nitrate—met the selection criterion of having maximum relative-concentrations greater than 0.5 (moderate or high) in the grid-based aquifer-scale proportions (fig. 9) and are listed in table 4. Figure 10 shows inorganic constituents that had relative-concentrations greater than 1.0 in one or more of the grid wells. Inorganic constituents having human-health benchmarks, as a group (nutrients, trace elements, and radioactive constituents), had high relative-concentrations in 9.1 percent of the primary aquifers, moderate relative-concentrations in 33.3 percent, and low relative-concentrations in 57.6 percent (table 7). To illustrate the spatial distributions of inorganic constituent concentrations, figures 11A–11B are maps showing inorganic data for USGS-grid wells and CDPH wells from the period May 1, 2002–July 20, 2005.

Iron and Other Trace Elements

Trace elements, as a class, had high relative-concentrations (for one or more constituents) in 1.3 percent of the primary aquifers, moderate values in 9.4 percent, and low values in 89.3 percent (table 7). The percentage of the primary aquifers with high and moderate relative-concentrations for the individual constituents is shown in table 4. Chromium had a spatially weighted high relative-concentration in 1.3 percent of the primary aquifers (table 4). The spatially weighted approach includes data from a larger number of wells than the grid-based approach, and therefore is more likely to include constituents present at high concentrations in small proportions of the primary aquifers. Among trace elements with SMCLs, iron had a high relative-concentration in 3.0 percent of the primary aquifers (table 4, fig. 11A).

The trace elements aluminum, lead, and vanadium had high relative-concentrations in at least one well reported in the CDPH database during the current period of study (May 1, 2002–April 30, 2005), but these high values were not the most recently reported values used for calculating aquifer-scale proportions.

The trace elements antimony, arsenic, cadmium, hexavalent chromium, manganese, mercury, molybdenum, and nickel had high relative-concentrations in at least one well reported in the CDPH database before May 1, 2002 (table 6), but not during the current period of study (May 1, 2002–April 30, 2005). These high values represent historical values rather than current values. Most of the historically high arsenic sites are located in south San Gabriel just to the east of the Merced Hills (fig. 2), while the historically high manganese sites are distributed throughout the San Fernando and San Gabriel study areas. Arsenic was detected at high relative-concentrations in 13 wells out of 517 with data; manganese was detected at high relative-concentrations in 23 wells out of 523 with data.

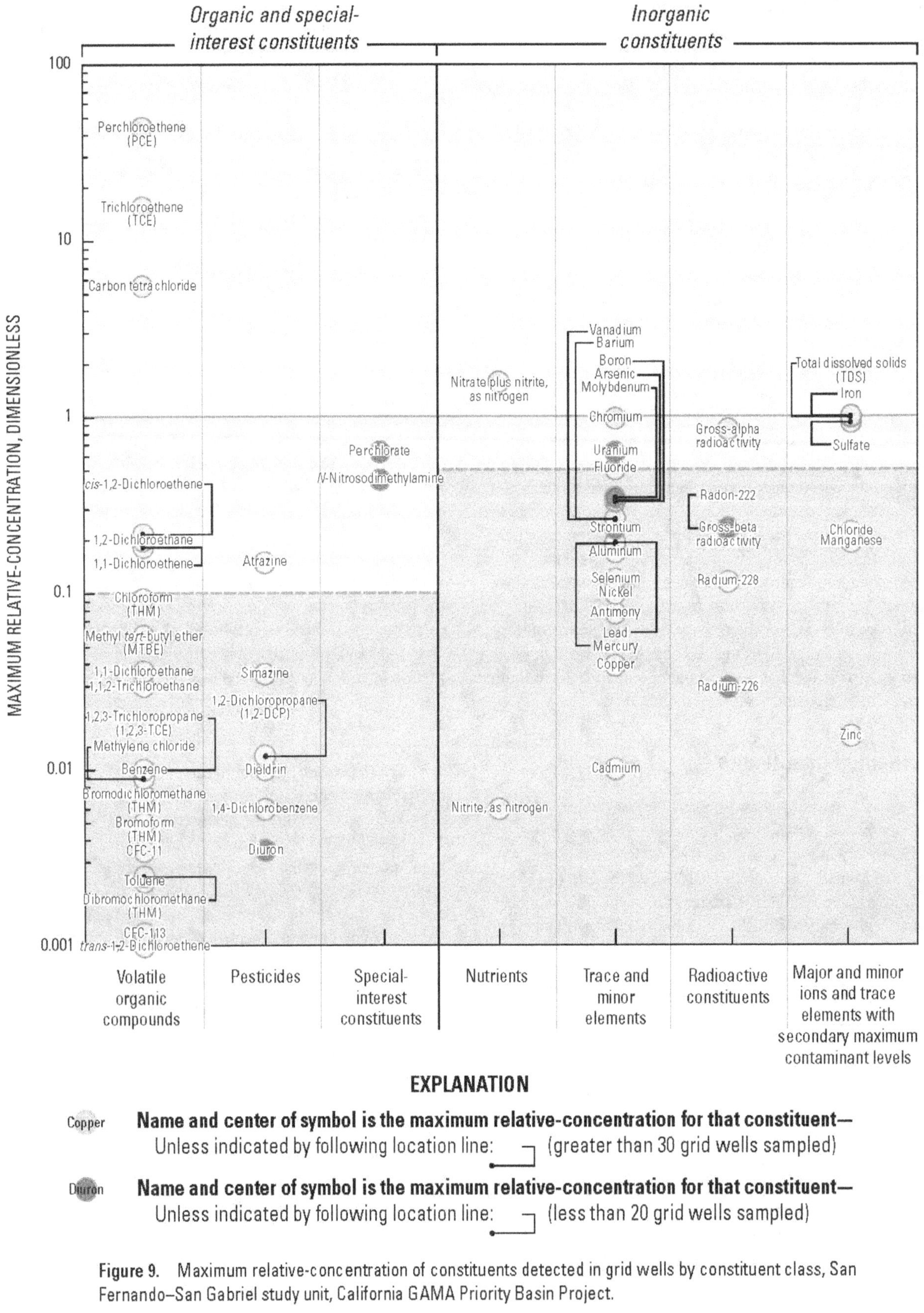

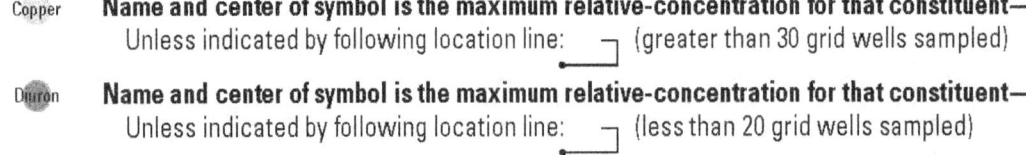

Figure 9.   Maximum relative-concentration of constituents detected in grid wells by constituent class, San Fernando–San Gabriel study unit, California GAMA Priority Basin Project.

Table 7. Aquifer-scale proportions for constituent classes, San Fernando–San Gabriel study unit, California GAMA Priority Basin Project.

[Aquifer-scale proportion: High, concentrations greater than benchmark; Moderate, concentrations less than benchmark and greater than or equal to 0.1 of benchmark for organic constituents or 0.5 of benchmark for inorganic constituents; Low, concentrations less than 0.1 of benchmark for organic constituents or 0.5 of benchmark for inorganic constituents; VOC, volatile organic compounds; SMCL, secondary maximum contaminant level; values are grid based except where footnoted]

Constituent class	Aquifer-scale proportion		
	Low relative-concentration (percent)	Moderate relative-concentration (percent)	High relative-concentration (percent)
Organic constituents with human-health benchmarks			
Solvents	41.8	40.0	[1] 18.2
Trihalomethanes	96.8	[1] 3.2	0.0
Other VOCs	91.4	5.7	[1] 2.9
Herbicides	94.3	5.7	0.0
Total for organic constituents with human-health benchmarks	38.9	42.9	[1] 18.2
Constituents of special interest			
Perchlorate	76.8	[1] 12.0	[1] 11.2
N-Nitrosodimethylamine (NDMA)	86.9	[1] 7.9	[1] 5.2
Inorganic constituents with human-health benchmarks			
Minor elements	96.2	3.1	[1] 0.7
Trace elements	89.3	9.4	[1] 1.3
Nutrients	64.7	26.5	8.8
Radioactive constituents	98.7	[1] 1.0	[1] 0.3
Total for inorganic constituents with human-health benchmarks	57.6	33.3	9.1
Inorganic constituents with SMCL benchmarks			
Total dissolved solids (TDS) (SMCL)	81.6	18.2	[1] 0.2
Sulfate (SMCL)	96.9	3.1	0.0
Iron (SMCL)	97.0	0.0	3.0
Total for inorganic constituents with SMCL benchmarks	78.6	18.2	3.2

[1] Spatially weighted value.

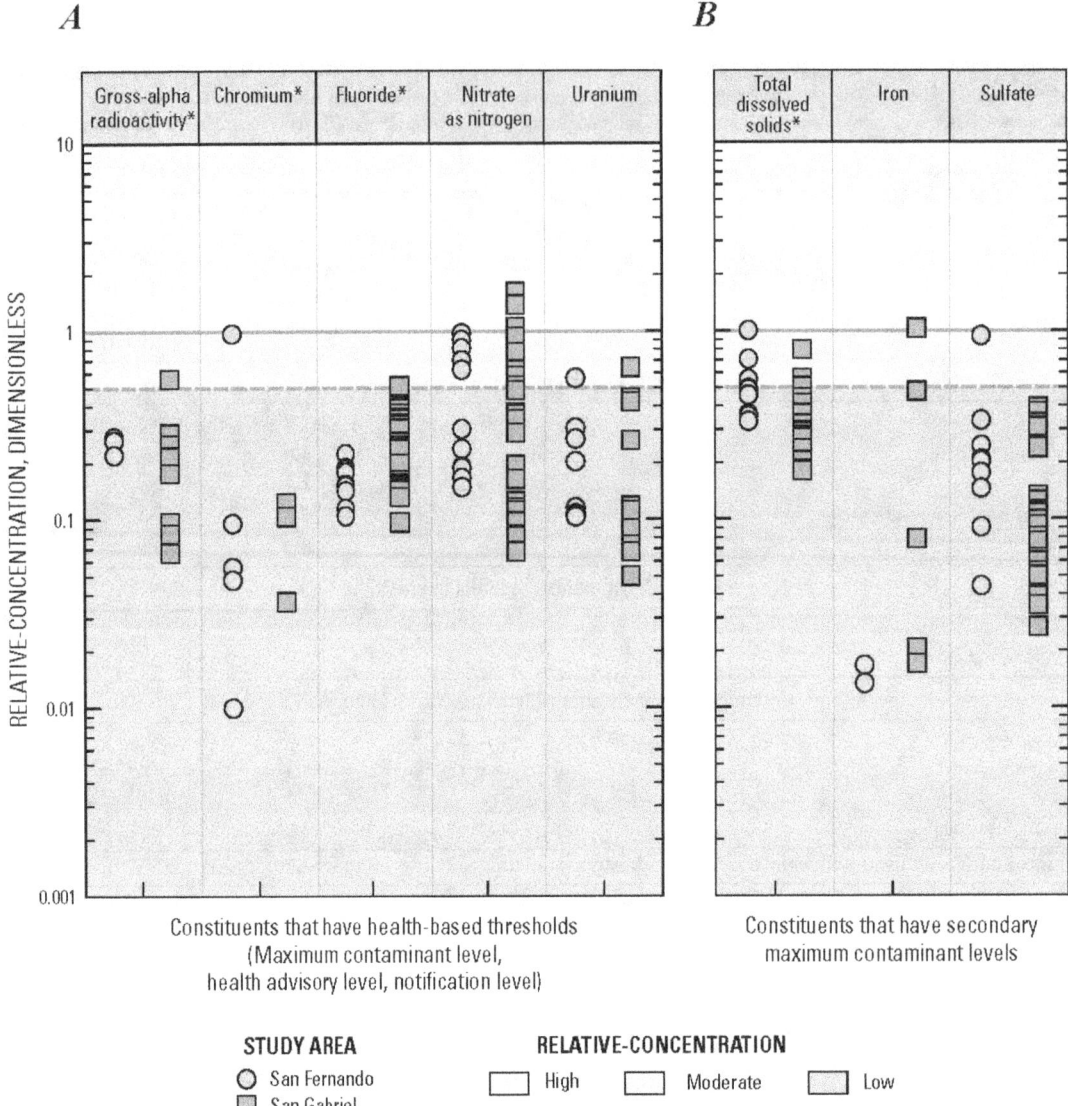

Figure 10.   Relative-concentrations of (*A*) gross-alpha radioactivity, chromium, fluoride, nitrate, and uranium with health-based benchmarks and relative-concentrations of (*B*) total dissolved solids, iron, and sulfate with aesthetic benchmarks in USGS and CDPH grid wells San Fernando–San study unit, California GAMA Priority Basin Project.

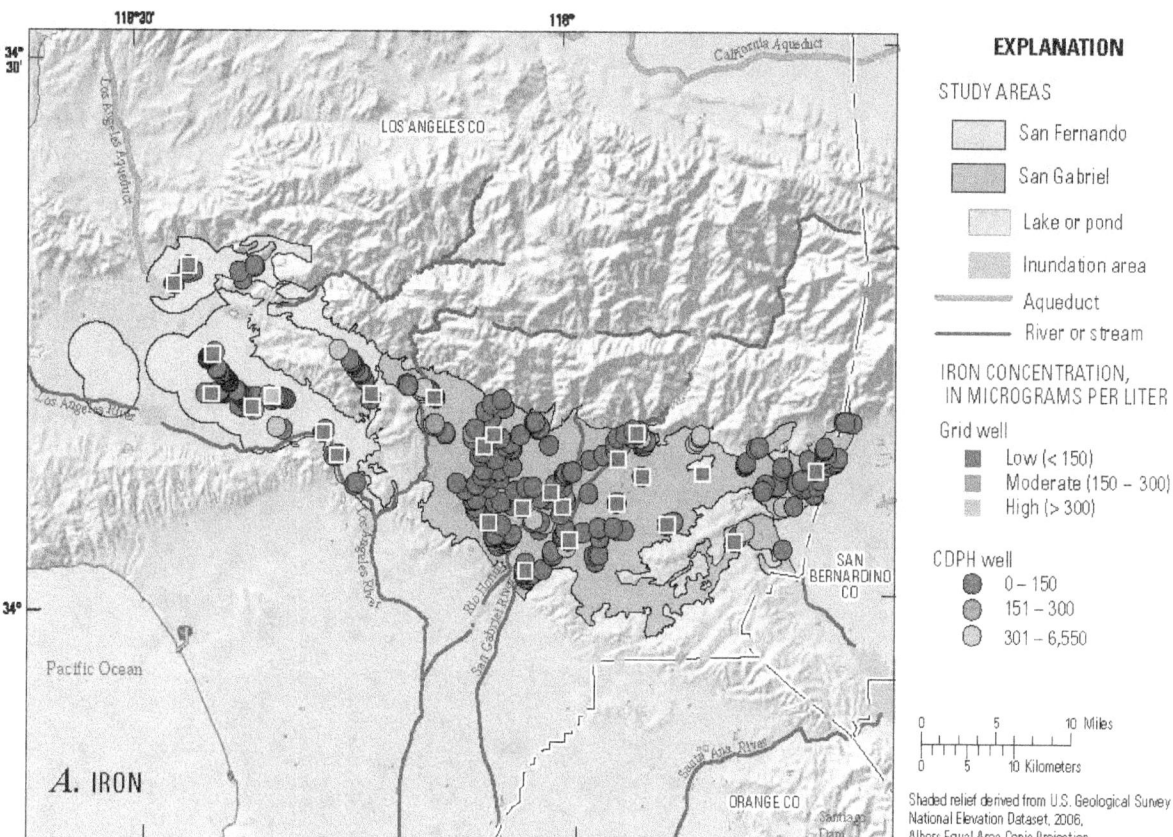

Figure 11.   Relative-concentrations of selected inorganic constituents in the U.S. Geological Survey (USGS) grid and USGS-understanding wells and California Department of Public Health (CDPH) wells (data from the period May 1, 2002–April 30, 2005), San Fernando–San Gabriel study unit, California GAMA Priority Basin Project.

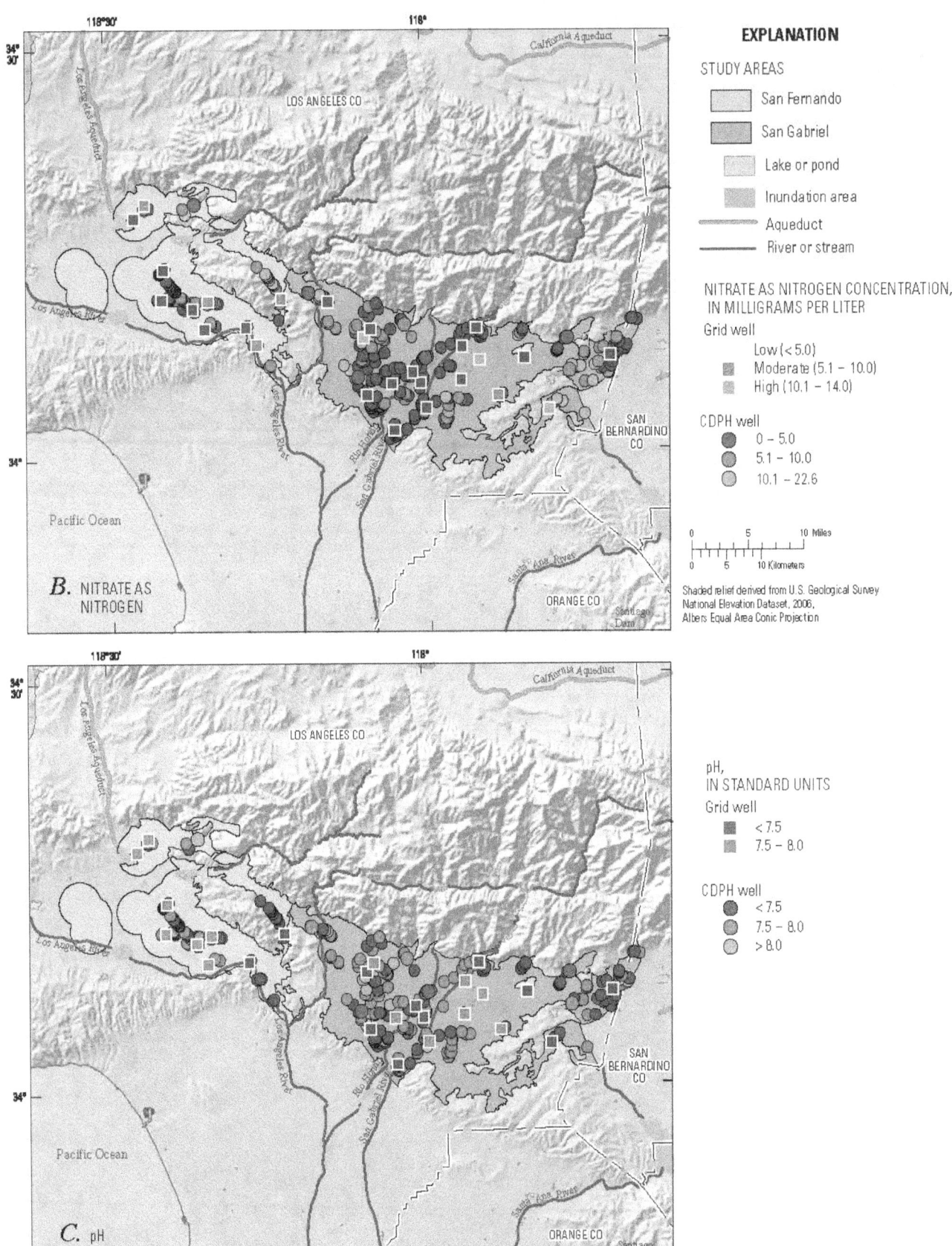

Figure 11.—Continued

Radioactive Constituents

The high relative-concentrations of radioactive constituents was 0.3 percent in the primary aquifers of the FG study unit based on the spatially weighted approach (table 7), reflecting the detection of gross-alpha radioactivity (30-day count). Gross-alpha radioactivity was detected at moderate relative-concentration in 1.0 percent of the primary aquifers (table 4).

Combined radium-226 plus radium-228 had high relative-concentrations in at least one well reported in the CDPH database during the current period of study (May 1, 2002– April 30, 2005), but this high value was not the most recently reported value used for calculating aquifer-scale proportions. In addition, radium-226, radium-228, and gross-beta radioactivity were detected at high relative-concentration in at least one well reported in the CDPH database before 2002, but not during the current period of study; these high values represent historical values rather than current values (table 6).

Nitrate and Other Nutrients

Nutrients as a class had high relative-concentrations in 8.8 percent of the primary aquifers and moderate relative-concentrations in 26.5 percent (table 7) resulting from the detection of nitrate plus nitrite, as nitrogen (hereinafter referred to as nitrate) (table 4). Nitrate was detected at high relative-concentrations in 8.8 percent of grid wells, and at moderate relative-concentration in 26.5 percent of the grid wells (table 4). Nitrate relative-concentration distribution is shown in figure 11B.

Nitrite was detected at high relative-concentration in at least one well reported in the CDPH database before 2002 (table 6), but not during the current period of study; these high values represent historical values rather than current values.

Major and Minor Ions

The major ions chloride and sulfate, and TDS have upper SMCL-CA benchmarks based on aesthetic properties. The minor ion fluoride has an MCL-CA, and the remaining seven major or minor ions do not have benchmarks.

Fluoride was detected at high relative-concentrations in 0.7 percent of the primary aquifers, on the basis of the spatially weighted approach, and at moderate relative-concentrations in 3.1 percent, on the basis of the grid-based approach (table 4). Among major and minor ions with SMCLs, TDS was detected at high relative-concentration in 0.2 percent of the primary aquifers, based on the spatially weighted approach, and was detected at moderate relative-concentration in 18.2 percent of the primary aquifers, based on the grid-based approach. Sulfate was not detected at high relative-concentrations, and was detected at moderate relative-concentration in 3.1 percent of the primary aquifers (table 4).

Organic Constituents

The organic compounds are organized by constituent class, including three classes of volatile organic compounds (VOCs) and two classes of pesticides. VOCs may be in paints, solvents, fuels, and refrigerants; VOCs can be byproducts of water disinfection and are characterized by a volatile nature, or tendency to evaporate. In this report, VOCs are classified into three categories: (1) solvents, (2) trihalomethanes, and (3) other VOCs (including organic synthesis reagents, refrigerants, and gasoline additives). Pesticides are used to control weeds, fungi, or insects in agricultural and urban settings. In this report, pesticides are classified as herbicides or insecticides (including fumigants). One or more organic constituents were found in all of the 35 grid wells sampled in the FG study unit. Sixty-three organic compounds were detected of the 208 analyzed for in all wells, and 45 out of these 63 organic compounds have human-health benchmarks (table 5).

Carbon tetrachloride, PCE, and TCE were detected in at least one grid well at high relative-concentrations. Four additional organic compounds were detected at high concentrations in the FG study unit, but not in grid wells: 1,1-dichloroethane, 1,2-dichloroethane, cis-1,2-dichloroethene, and 1,1-dichloroethene. The proportion of the aquifer that had high relative-concentrations of organic constituents was 18.2 percent (table 7). Eight organic constituents had maximum relative-concentrations greater than 0.1 (moderate) (figs. 12 and 13). The proportion of the aquifer having moderate relative-concentrations of organic constituents was 42.9 percent (table 7).

Chloroform, PCE, simazine, atrazine, and TCE were detected in more than 50 percent of the grid wells. Bromodichloromethane, cis-1,2 dichloroethene, 1,1-dichloroethane, carbon tetrachloride, and 1,1-dichloroethene were detected in more than 30 percent of the grid wells. MTBE, prometon, and diuron were detected in more than 20 percent of the grid wells, and CFC-12, bromacil, carbon disulfide, 1,1,1-trichloroethane (TCA), CFC-113, tebuthiuron, dibromochloromethane, and CFC-11 were detected in more than 10 percent of the grid wells. However, diuron and bromacil were sampled in a subset of only 11 wells. The individual constituents that were not detected in the FG study unit are listed in Land and Belitz (2008). Pharmaceutical compounds were not detected at concentrations greater than or equal to method detection limits in the FG study unit. Fram and Belitz (2011) present all results for pharmaceutical compounds in groundwater samples collected for the first 28 GAMA-PBP study units (May 2004 through March 2010), including the FG study unit.

EXPLANATION

Atrazine **Name and center of symbol is the maximum relative concentration for that constituent—**
Unless indicated by following location line: (greater than 35 grid wells sampled)

Diuron **Name and center of symbol is the maximum relative concentration for that constituent—**
Unless indicated by following location line: (11 grid wells sampled)

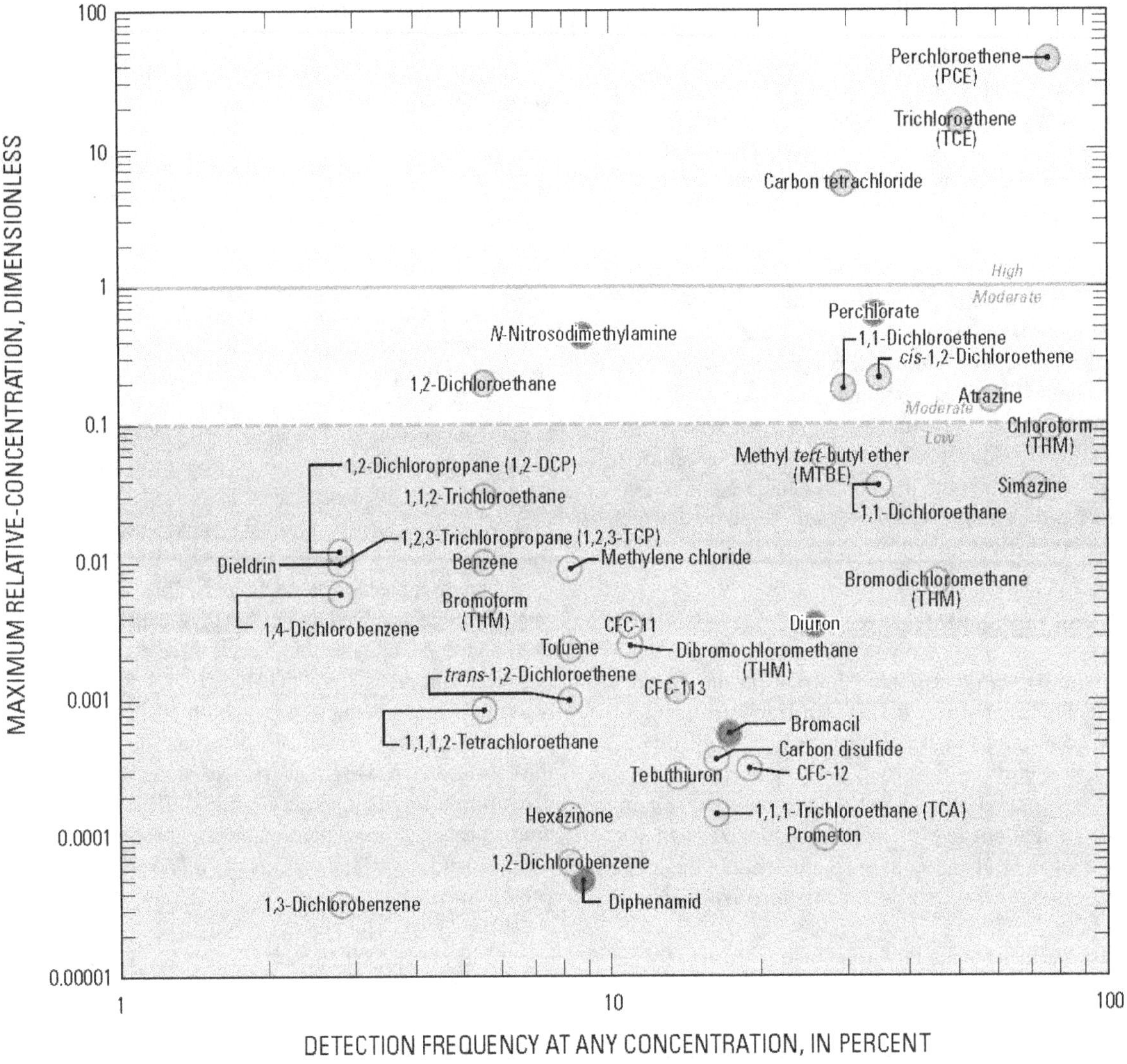

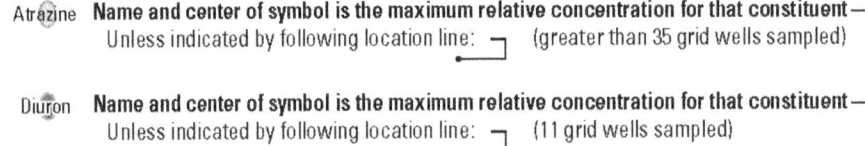

Figure 12. Detection frequency and maximum relative-concentration of organic and special-interest constituents detected in USGS-grid wells in the San Fernando–San Gabriel study unit, California GAMA Priority Basin Project.

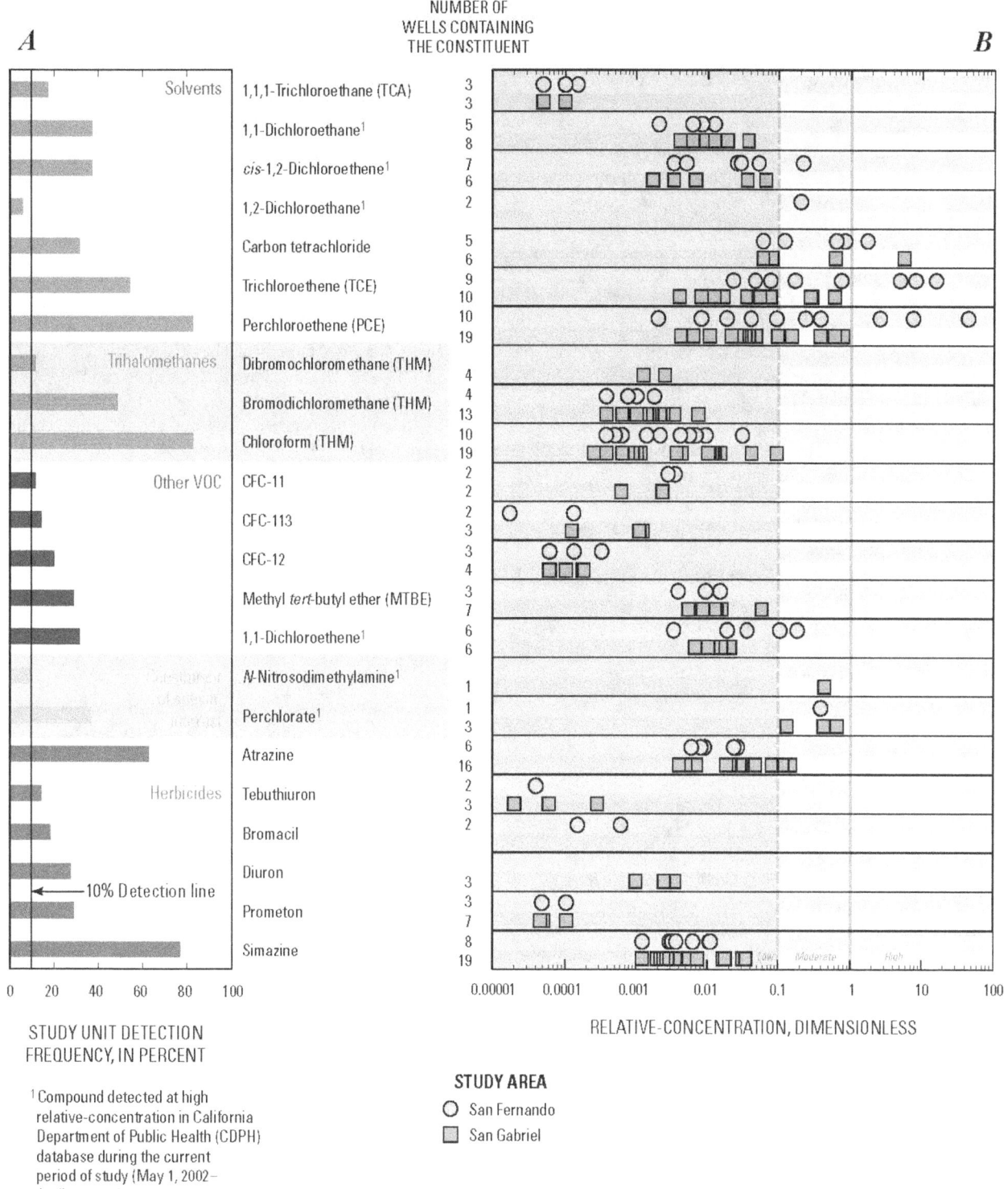

Figure 13. (*A*) Detection frequency and (*B*) relative-concentrations of selected organic and special-interest constituents in grid wells in the San Fernando–San Gabriel study unit, California GAMA Priority Basin Project, May–August 2005.

Carbon Tetrachloride, Perchloroethene, Trichloroethene, and Other Solvents

Solvents are used for various industrial, commercial, and domestic purposes. Solvents, as a class of VOCs, had high relative-concentrations in 18.2 percent of the primary aquifers, and moderate relative-concentrations in 40.0 percent (table 7). The solvent TCE was detected at a high aquifer-scale proportion in 14.8 percent of the primary aquifers (table 4). The solvent PCE was detected at a high aquifer-scale proportion in 11.2 percent of the primary aquifers. The solvent carbon tetrachloride was detected at a high aquifer-scale proportion in 6.5 percent of the primary aquifers. The high aquifer-scale proportions for TCE, PCE, and carbon tetrachloride were calculated using the spatially weighted approach. The following solvents had spatially weighted high aquifer-scale proportions: 1,1-dichloroethane (0.5 percent), 1,2-dichloroethane (1.9 percent), and cis-1,2-dichloroethene (1.0 percent). Six solvents—carbon tetrachloride, PCE, TCE, 1,1-dichloroethane, cis-1,2-dichloroethene, and TCA—were detected in more than 10 percent of the grid wells sampled (fig. 12). The spatial distribution of these solvents, based on data for USGS-grid wells and CDPH wells for the period May 1, 2002 to July 1, 2005, are shown on maps (figs. 14A-G)

Historically high values for the solvents 1,1,1-trichloroethane, 1,1,2-trichloroethane, 1,2,3-trichloropropane, benzene, dichloromethane, and trans-1,2-dichloroethene were recorded in the CDPH database for the period before May 1, 2002 (table 6) but not during the current period of study.

Trihalomethanes

The class "trihalomethanes" (THMs) had constituents with moderate relative-concentrations in the primary aquifers, on the basis of the spatially weighted approach, but no concentrations were above health-based benchmarks (table 7). Chloroform, bromodichloromethane, and dibromochloromethane were detected at 2.1 percent, 0.7 percent, and 0.5 percent moderate relative-concentrations, respectively, based on the spatially weighted approach (table 4). These three compounds also were detected in more than 10 percent of the grid wells (fig. 12). The total THMs (sum of the individual THMs) were calculated to be in 3.3 percent of aquifers at a moderate relative-concentration, based on the spatially weighted approach. Relative-concentrations of total THMs are shown in figure 14H for USGS-grid wells and CDPH wells during May 1, 2002–July 20, 2005.

1,1-Dichloroethene and Other Volatile Organic Compounds

Other VOCs as a class, which includes organic synthesis reagents, refrigerants, and gasoline additives, were detected at high relative-concentrations (for one or more constituents) in 2.9 percent of the primary aquifers (based on the spatially weighted approach), and at moderate relative-concentrations in 5.7 percent (table 7). The high aquifer-scale proportion reflects the relative-concentration of the organic synthesis reagent 1,1-dichloroethene (table 4). The refrigerant CFC-11 (and total CFCs) and the gasoline additive MTBE, also included in the class "other VOCs," were detected at moderate concentrations in less than 1 percent of the primary aquifers, based on the spatially weighted approach. The refrigerants CFC-11, CFC-12, CFC-113, 1,1-dichloroethene, carbon disulfide, and MTBE were detected in more than 10 percent of the grid wells (fig. 12). Relative-concentrations of total chlorofluorocarbons (CFCs) (sum of CFCs), 1,1-dichloroethene, and MTBE are shown in figures 14I–K for USGS-grid wells and CDPH wells during May 1, 2002–July 20, 2005.

Historically high values for 1,4-dioxane, bis(2-ethylhexyl)phthalate (DEHP), and vinyl chloride were recorded in the CDPH database for the period before May 1, 2002, but not during the current period of study (table 6).

Herbicides

Herbicides are commonly used to control weeds in agricultural, urban, and rural areas. No herbicides were detected at high relative-concentrations; however, moderate relative-concentrations of the herbicide atrazine were detected in samples from the FG study unit (figs. 12, 13). Atrazine, prometon, simazine, and tebuthiuron were detected in more than 10 percent of the grid wells (figs. 14L–O). Simazine was among the most commonly detected herbicides in groundwater in major aquifers across the United States (Gilliom and others, 2006). Simazine also was among the most frequently detected triazine herbicides in groundwater in California (Troiano and others, 2001), and is used on rights-of-way for weed control (Domagalski and Dubrovsky, 1991).

Herbicides, as a class, were detected at moderate relative-concentrations in 5.7 percent of the primary aquifers, which reflects the detection of atrazine (tables 4, 7).

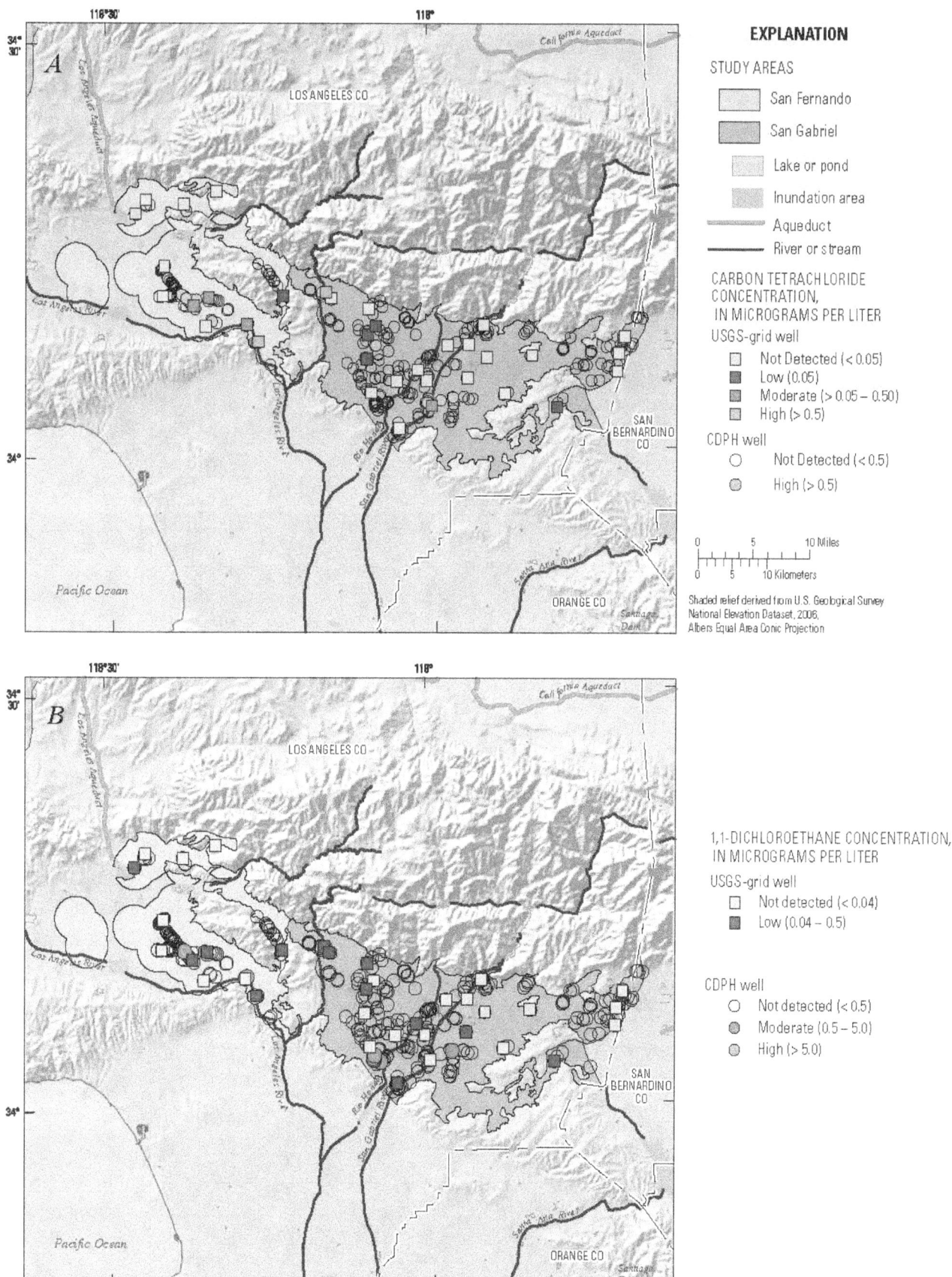

Figure 14. Relative concentrations for selected organic constituents, San Fernando–San Gabriel study unit, California GAMA Priority Basin Project.

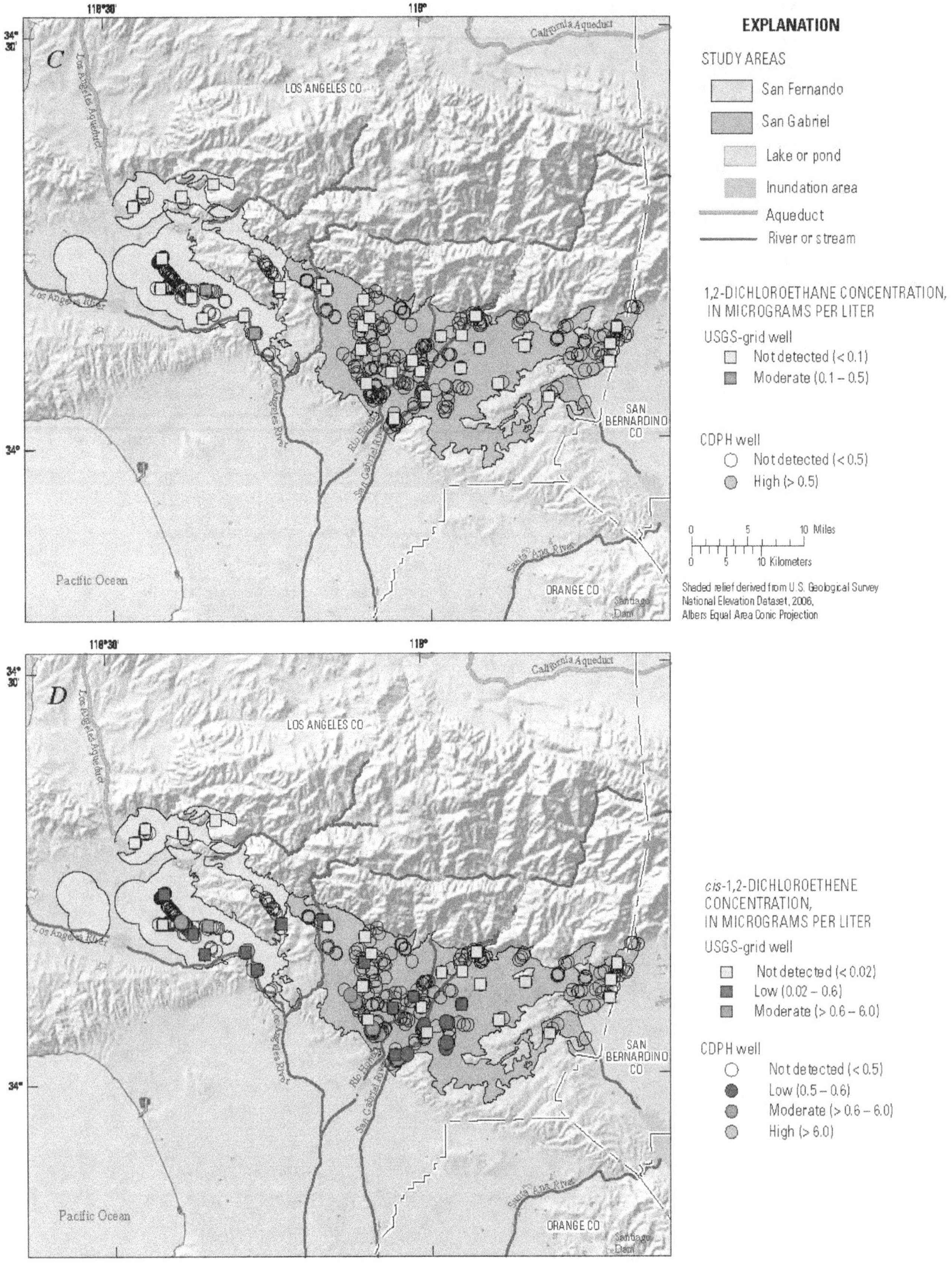
118°30′
118°
34° 30′
34°
C
California Aqueduct
Los Angeles Aqueduct
LOS ANGELES CO
Los Angeles River
Rio Hondo
San Gabriel River
SAN BERNARDINO CO
Santa Ana River
ORANGE CO
Santiago Dam
Pacific Ocean
EXPLANATION
STUDY AREAS
San Fernando
San Gabriel
Lake or pond
Inundation area
Aqueduct
River or stream
1,2-DICHLOROETHANE CONCENTRATION, IN MICROGRAMS PER LITER
USGS-grid well
Not detected (< 0.1)
Moderate (0.1 – 0.5)
CDPH well
Not detected (< 0.5)
High (> 0.5)
0 5 10 Miles
0 5 10 Kilometers
Shaded relief derived from U.S. Geological Survey National Elevation Dataset, 2006, Albers Equal Area Conic Projection
118°30′
118°
34° 30′
34°
D
California Aqueduct
Los Angeles Aqueduct
LOS ANGELES CO
Los Angeles River
Rio Hondo
San Gabriel River
SAN BERNARDINO CO
Santa Ana River
ORANGE CO
Santiago Dam
Pacific Ocean
cis-1,2-DICHLOROETHENE CONCENTRATION, IN MICROGRAMS PER LITER
USGS-grid well
Not detected (< 0.02)
Low (0.02 – 0.6)
Moderate (> 0.6 – 6.0)
CDPH well
Not detected (< 0.5)
Low (0.5 – 0.6)
Moderate (> 0.6 – 6.0)
High (> 6.0)

Figure 14.—Continued

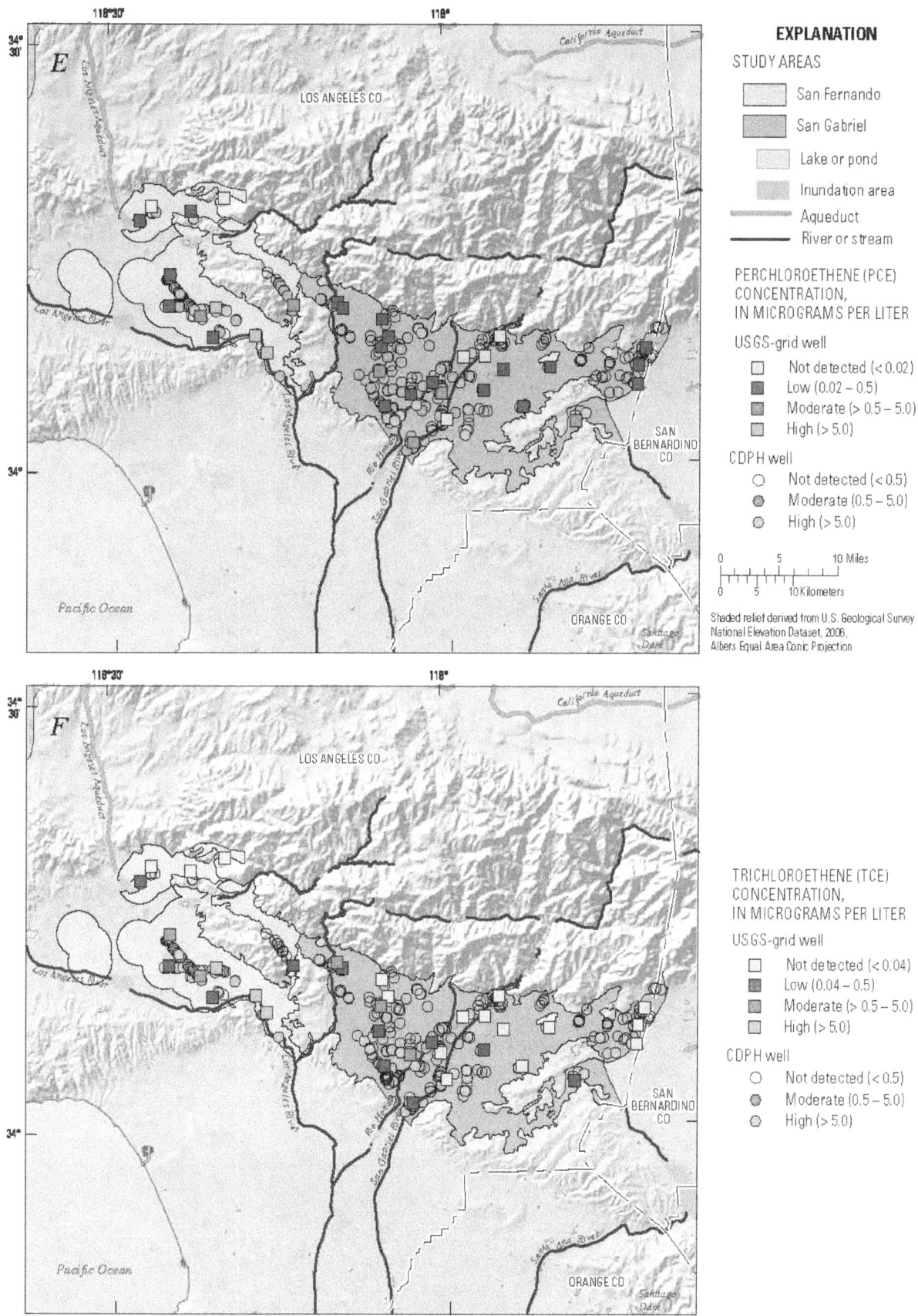

EXPLANATION

STUDY AREAS

San Fernando

San Gabriel

Lake or pond

Inundation area

Aqueduct

River or stream

PERCHLOROETHENE (PCE) CONCENTRATION, IN MICROGRAMS PER LITER

USGS-grid well

Not detected (< 0.02)

Low (0.02 – 0.5)

Moderate (> 0.5 – 5.0)

High (> 5.0)

CDPH well

Not detected (< 0.5)

Moderate (0.5 – 5.0)

High (> 5.0)

0   5   10 Miles

0   5   10 Kilometers

Shaded relief derived from U.S. Geological Survey
National Elevation Dataset, 2006,
Albers Equal Area Conic Projection

TRICHLOROETHENE (TCE) CONCENTRATION, IN MICROGRAMS PER LITER

USGS-grid well

Not detected (< 0.04)

Low (0.04 – 0.5)

Moderate (> 0.5 – 5.0)

High (> 5.0)

CDPH well

Not detected (< 0.5)

Moderate (0.5 – 5.0)

High (> 5.0)

Figure 14.—Continued

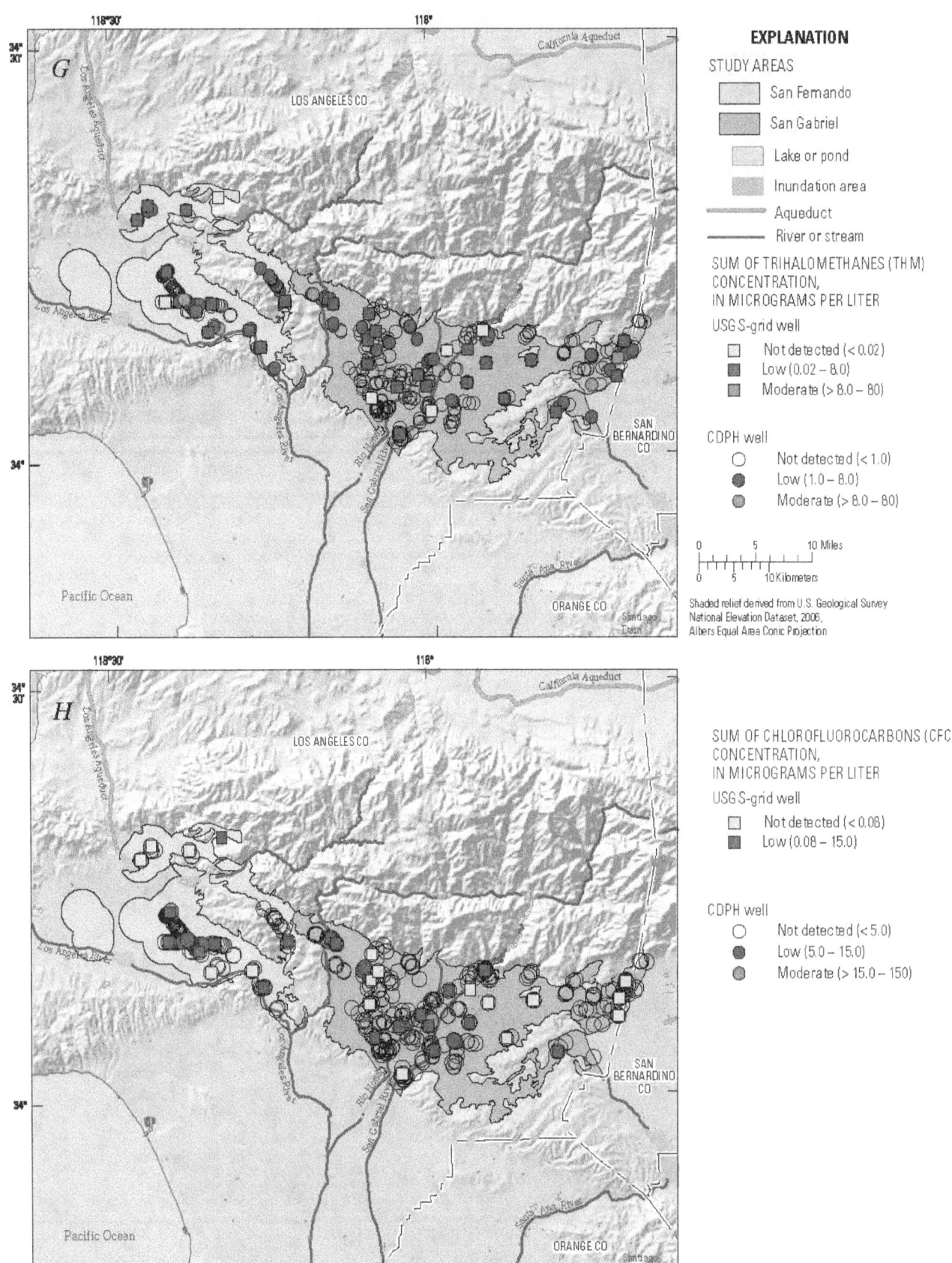

Figure 14.—Continued

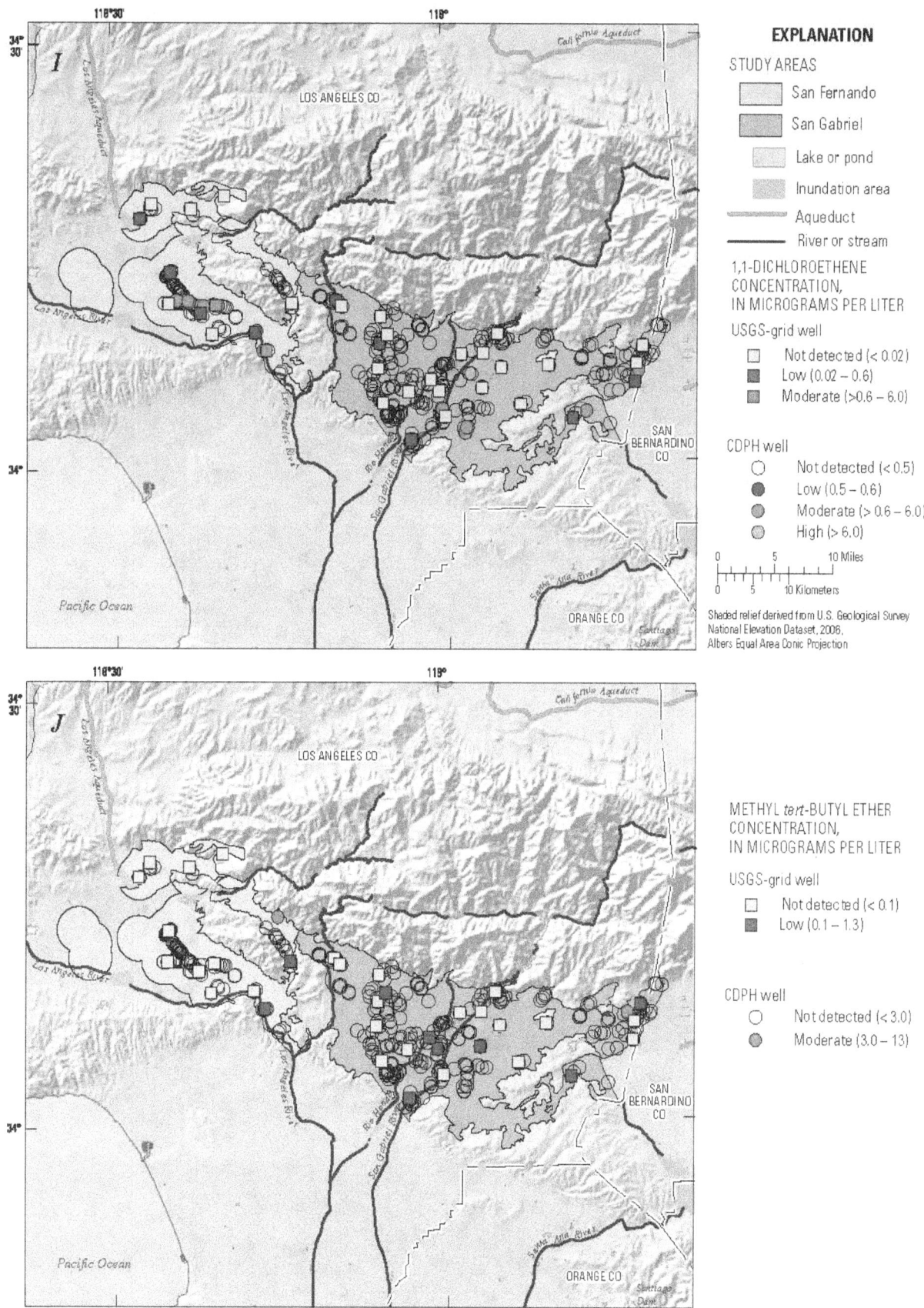

EXPLANATION

STUDY AREAS

San Fernando

San Gabriel

Lake or pond

Inundation area

Aqueduct

River or stream

1,1-DICHLOROETHENE
CONCENTRATION,
IN MICROGRAMS PER LITER

USGS-grid well

☐ Not detected (< 0.02)

◼ Low (0.02 – 0.6)

◼ Moderate (>0.6 – 6.0)

CDPH well

○ Not detected (< 0.5)

● Low (0.5 – 0.6)

● Moderate (> 0.6 – 6.0)

● High (> 6.0)

0   5   10 Miles

0   5   10 Kilometers

Shaded relief derived from U.S. Geological Survey
National Elevation Dataset, 2006,
Albers Equal Area Conic Projection

METHYL *tert*-BUTYL ETHER
CONCENTRATION,
IN MICROGRAMS PER LITER

USGS-grid well

☐ Not detected (< 0.1)

◼ Low (0.1 – 1.3)

CDPH well

○ Not detected (< 3.0)

● Moderate (3.0 – 13)

Figure 14.—Continued

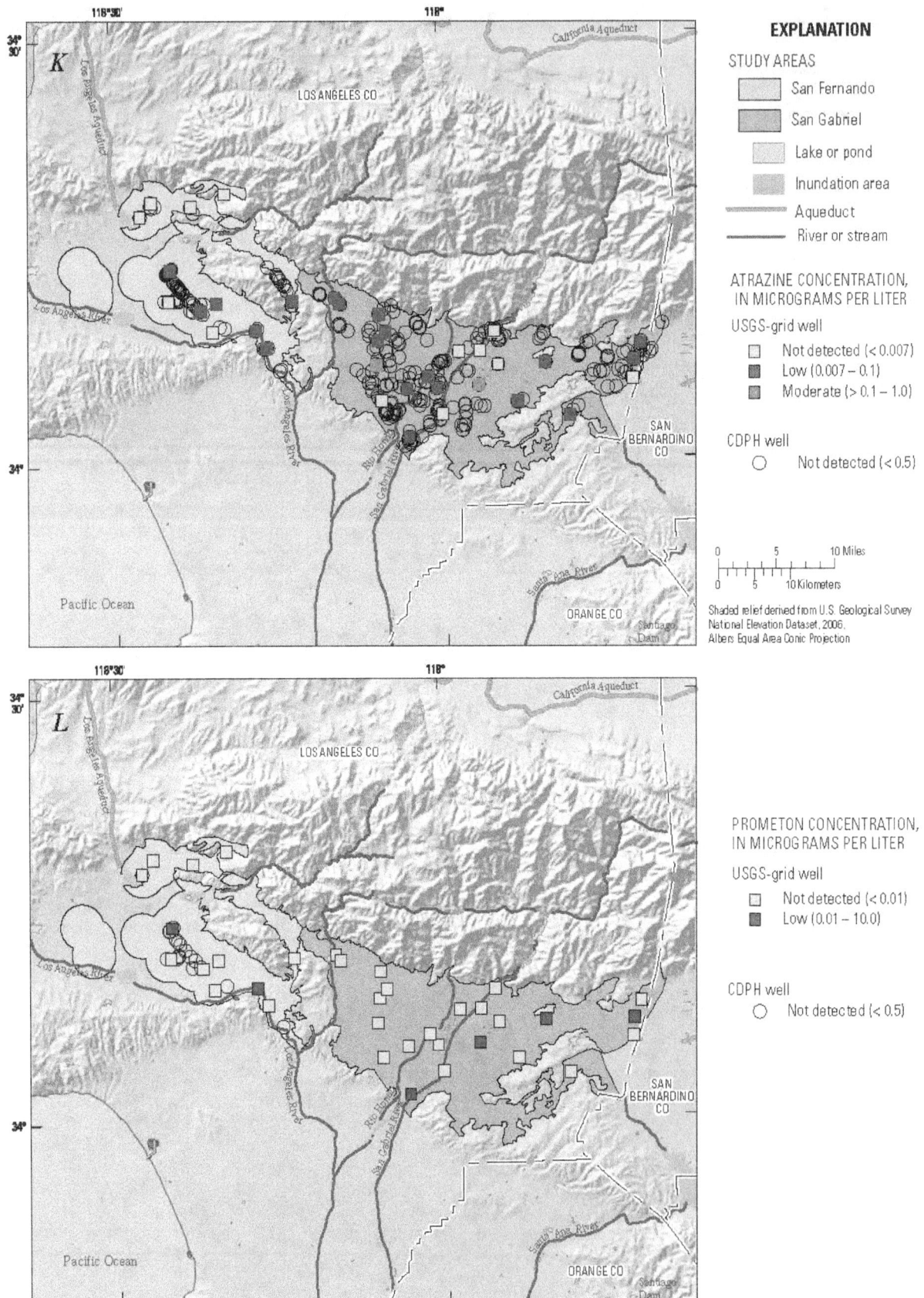

Figure 14.—Continued

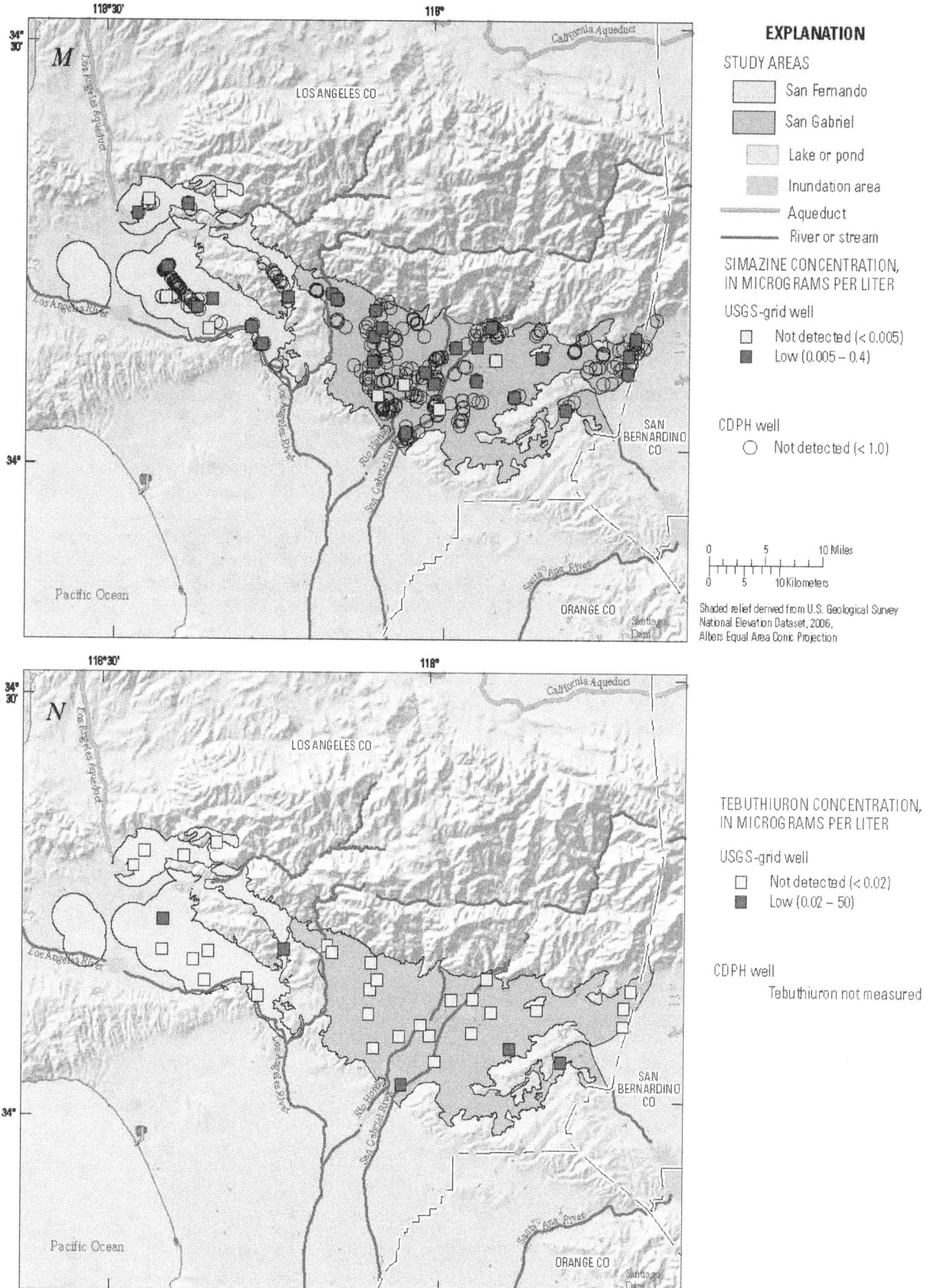

EXPLANATION

STUDY AREAS

San Fernando

San Gabriel

Lake or pond

Inundation area

Aqueduct

River or stream

SIMAZINE CONCENTRATION,
IN MICROGRAMS PER LITER

USGS-grid well

Not detected (< 0.005)

Low (0.005 – 0.4)

CDPH well

Not detected (< 1.0)

0   5   10 Miles

0   5   10 Kilometers

Shaded relief derived from U.S. Geological Survey
National Elevation Dataset, 2006,
Albers Equal Area Conic Projection

TEBUTHIURON CONCENTRATION,
IN MICROGRAMS PER LITER

USGS-grid well

Not detected (< 0.02)

Low (0.02 – 50)

CDPH well

Tebuthiuron not measured

Figure 14.—Continued

Insecticides

Two fumigants, 1,2-dichloropropane and 1,4-dichlorobenzene, were detected at low relative-concentrations in samples from USGS-grid wells (fig. 12). Dibromochloropropane (DBCP) had a high relative-concentration at one well in the CDPH database, but this high value was not the most recently reported value used for calculating aquifer-scale proportion (table 4).

Historically-high values for the insecticides heptachlor and lindane and the fumigant 1,2-dibromomethane (EDB) were recorded in the CDPH database for the period before May 1, 2002, but not during the current period of study (table 6).

Constituents Analyzed in Groundwater from a Subset of Wells

Data for 62 constituents, including polar pesticides, selected trace elements, selected radioactive constituents, and four constituents of special interest, were available for only a subset of grid wells (less than 20 wells) (Land and Belitz, 2008). Two organic constituents, diuron and bromacil, met the selection criteria of having detection frequencies greater than 10 percent in the samples collected. Diuron was detected in 3 of the 11 grid wells sampled (27 percent). Diuron has been used extensively in central California on orchards, particularly oranges, but has also been used for nonagricultural purposes such as weed control on roadways (Domagalski and Dubrovsky, 1991). Nationally, the detection frequency of diuron was less than 1 percent (Gilliom and others, 2006), but in California it has been detected in about 6 percent of well samples (Troiano and others, 2001). Another herbicide, bromacil, was detected in 2 of the 11 grid wells sampled (18 percent). Because diuron and bromacil were analyzed in less than one-half of the grid wells, the detection frequency may not be representative of the study unit.

Perchlorate, NDMA, and Other Special-Interest Constituents

Constituents of special interest analyzed for in the FG study unit at 11 USGS-grid wells were NDMA, 1,2,3-TCP, 1,4-dioxane, and perchlorate. These constituents were selected because they recently have been detected in drinking-water supplies, or are considered to have the potential to reach drinking-water supplies (California Department of Public Health, 2008a,b,c). Additional data were available from the CDPH database, and were used in combination with USGS GAMA data to calculate aquifer-scale proportions for perchlorate and NDMA. Perchlorate was detected at a spatially weighted high aquifer-scale proportion of 11.2 percent, and at a moderate aquifer-scale proportion of 12.0 percent. NDMA was detected at a spatially weighted high aquifer-scale proportion of 5.2 percent, and at a moderate aquifer-scale proportion of 7.9 percent (table 4; figs. 15A–B). 1,2,3-TCP was detected at low relative-concentrations in the primary aquifers (fig. 12), and 1,4-dioxane was not detected in the FG study unit.

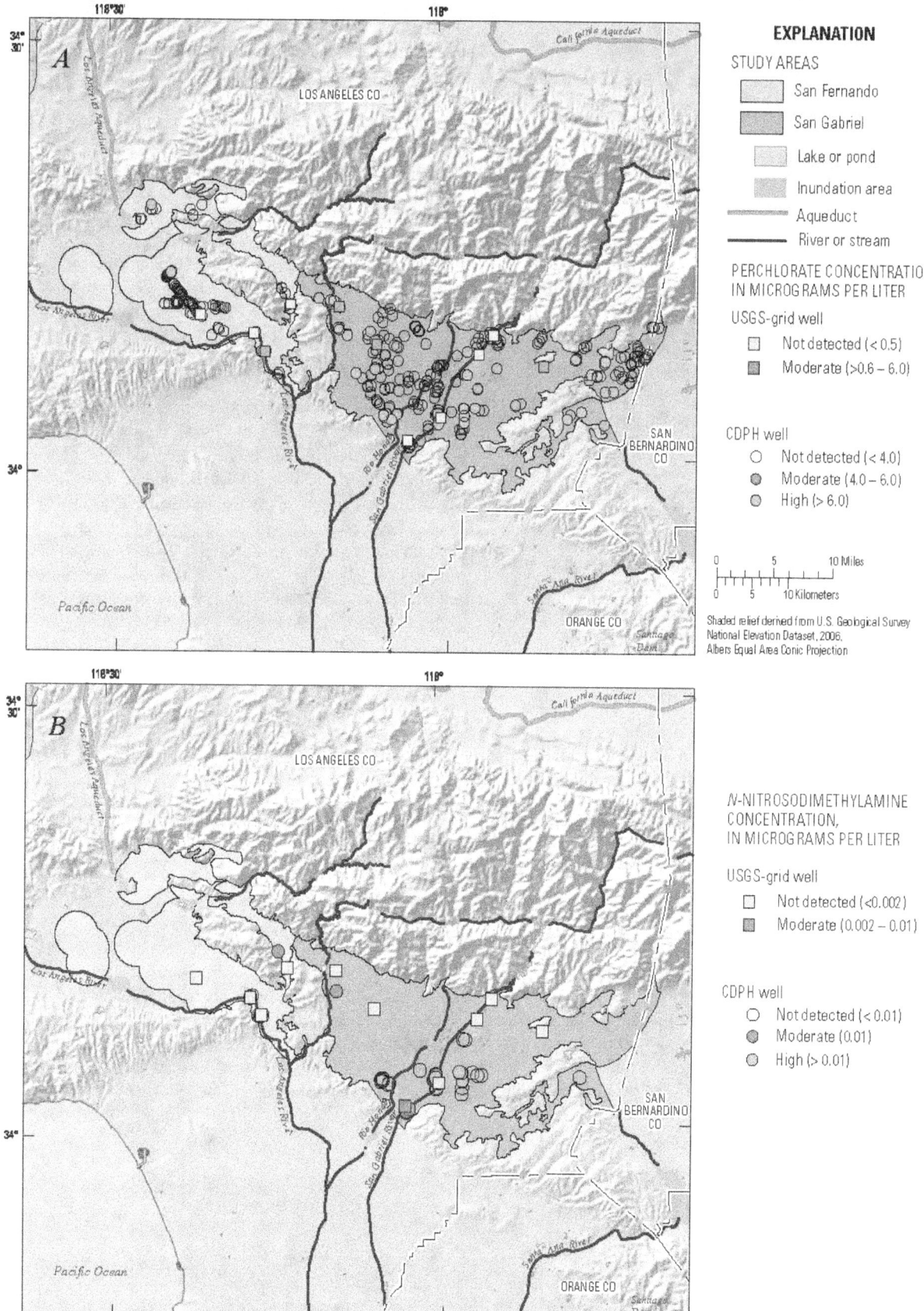

Figure 15.   Relative concentration of perchlorate and *N*-nitrosodimethylamine, San Fernando–San Gabriel study unit, California GAMA Priority Basin Project.

Summary

Groundwater quality in the approximately 460-square-mile San Fernando–San Gabriel (FG) study unit was investigated as part of the Priority Basin Project of the Groundwater Ambient Monitoring and Assessment (GAMA) Program. The GAMA FG study provides a spatially unbiased characterization of untreated groundwater quality in the primary aquifers. The assessment is based on water-quality and ancillary data collected in 2005 by the U.S. Geological Survey (USGS) from 52 wells, and on water-quality data from the California Department of Public Health (CDPH) database (during May 1, 2002–April 30, 2005).

The first component of this study, the status of the current quality of the groundwater resource, was assessed by using data from samples analyzed for volatile organic compounds (VOCs), pesticides, and naturally occurring inorganic constituents, such as major ions and trace elements. The status assessment characterizes the quality of groundwater resources in the primary aquifers of the FG study unit, not the treated drinking water delivered to consumers by water purveyors.

Relative-concentrations (sample concentration divided by the health- or aesthetic-based benchmark concentration) were used for evaluating groundwater quality for those constituents that have Federal and (or) California regulatory or non-regulatory benchmarks for drinking-water quality.

Aquifer-scale proportion was used as the primary metric for evaluating regional-scale groundwater quality. High aquifer-scale proportion is defined as the percentage of the primary aquifers with relative-concentration greater than 1.0 for a particular constituent or class of constituents; proportion is based on an areal rather than a volumetric basis. Moderate and low aquifer-scale proportions were defined as the percentage of the primary aquifers with moderate and low relative-concentrations, respectively. Two statistical approaches, grid-based and spatially weighted, were used to evaluate aquifer-scale proportions for individual constituents and classes of constituents. Grid-based and spatially weighted estimates were comparable in the FG study unit (within 90-percent confidence intervals). However, the spatially weighted approach was superior to the grid-based proportion when a constituent is high in a small fraction of the aquifer.

Inorganic constituents with human-health benchmarks were present at high relative-concentrations in 9.1 percent of the primary aquifers and moderate in 33.3 percent. The high aquifer-scale proportion of inorganic constituents primarily reflected high aquifer-scale proportions of nitrate plus nitrite (8.8 percent). The inorganic constituents with secondary maximum contaminant levels had relative-concentrations that were high in 3.2 percent of the primary aquifers, and moderate in 18.2 percent.

Relative-concentrations of organic constituents (one or more) were high in 18.2 percent and moderate in 42.9 percent of the primary aquifers. The high aquifer-scale proportion for organic constituents primarily reflected high aquifer-scale proportions of TCE (14.8 percent), PCE (11.2 percent), and carbon tetrachloride (6.5 percent), as determined by the spatially weighted method. Of the 212 organic and special-interest constituents analyzed, 66 constituents were detected. Chloroform, PCE, simazine, atrazine, and TCE were detected in more than 50 percent of the grid wells. Bromodichloromethane, cis-1,2-dichloroethene, 1,1-dichloroethane, perchlorate, carbon tetrachloride, and 1,1-dichloroethene were detected in more than 30 percent of the grid wells. MTBE, prometon, and diuron were detected in more than 20 percent of the grid wells, and CFC-12, bromacil, carbon disulfide, 1,1,1-trichloroethane (TCA), CFC-113, tebuthiuron, dibromochloromethane, and CFC-11 were detected in greater than 10 percent of the grid wells. Perchlorate, diuron, and bromacil were sampled in a subset of only 11 wells, not in all 35 grid wells. Perchlorate and NDMA were detected at high relative-concentrations in 11.2 percent and 5.2 percent of the primary aquifers, respectively. Pharmaceutical compounds were not detected at concentrations greater than or equal to method detection limits in the FG study unit.

Twenty-four constituents (12 organic and 12 inorganic constituents) were detected at high relative-concentrations prior to 2002, but were not detected in groundwater samples during the period of study (May 2002–August 2005); these constituents reflect historical conditions.

Acknowledgments

The authors thank the following cooperators for their support: the State Water Resources Control Board, LLNL, CDPH, and CDWR. We especially thank the cooperating well owners and water purveyors for their generosity in allowing the USGS to collect samples from their wells. Funding for this work was provided by State of California bonds authorized by Proposition 50 and administered by the State Water Board.

References

Aeschbach-Hertig, W., Peeters, F., Beyerle, U., and Kipfer, R., 1999, Interpretation of dissolved atmospheric noble gases in natural waters: Water Resources Research, v. 35, no. 9, p. 2779–2792.

Aeschbach-Hertig, W., Peeters, F., Beyerle, U., and Kipfer, R., 2000, Paleotemperature reconstruction from noble gases in groundwater taking into account equilibration with entrapped air: Nature, v. 405, p. 1040–1044.

Andrews, J.N., 1985, The isotopic composition of radiogenic helium and its use to study groundwater movement in confined aquifers: Chemical Geology, v. 49, p. 339–351.

Andrews, J.N., and Lee, D.J., 1979, Inert gases in groundwater from the Bunter Sandstone of England as indicators of age and paleoclimatic trends: Journal of Hydrology, v. 41, p. 233–252.

Belitz, Kenneth, Dubrovsky, N.M., Burow, K.R., Jurgens, B.C., and Johnson, T., 2003, Framework for a groundwater quality monitoring and assessment program for California: U.S. Geological Survey Water-Resources Investigations Report 03–4166, 28 p. (Also available at http://pubs.usgs. gov/wri/wri034166/.)

Belitz, K., Jurgens, B., Landon, M.K., Fram, M.S., and Johnson, T., 2010, Estimation of aquifer-scale proportion using equal-area grids—Assessment of regional-scale groundwater quality: Water Resources Research, v. 46, W11550, 14 p., doi:10.1029/2010WR009321. (Also available at http://www.agu.org/pubs/ crossref/2010/2010WR009321.shtml.)

Brown, L.D., Cai, T.T., and DasGupta, A., 2001, Interval estimation for a binomial proportion: Statistical Science, v. 16, no. 2, p. 101–117. (Also available at http://www.jstor. org/stable/2676784.)

California Department of Public Health, 2008a, Perchlorate in drinking water: California Department of Public Health website, accessed October 17, 2009, at http://www.cdph. ca.gov/certlic/drinkingwater/Pages/Perchlorate.aspx.

California Department of Public Health, 2008b, California drinking water—NDMA-related activities: California Department of Public Health website, accessed October 17, 2009, at http://www.cdph.ca.gov/certlic/drinkingwater/ Pages/NDMA.aspx.

California Department of Public Health, 2008c, 1,2,3-Trichloropropane: California Department of Public Health website, accessed October 17, 2009, at http://www. cdph.ca.gov/certlic/drinkingwater/Pages/123TCP.aspx.

California Department of Water Resources, 2003, California's groundwater: California Department of Water Resources Bulletin 118, 246 p. (Also available at http://www.water. ca.gov/groundwater/bulletin118/update2003.cfm.)

California Department of Water Resources, 2005a, California Department of Water Resources Bulletin 118, individual basin descriptions, San Fernando Valley groundwater basin.

California Department of Water Resources, 2005b, California Department of Water Resources Bulletin 118, individual basin descriptions, San Gabriel Valley groundwater basin.

California Department of Water Resources, 2005c, California Department of Water Resources Bulletin 118, individual basin descriptions, Raymond groundwater basin.

California State Water Resources Control Board, 2003, A comprehensive groundwater quality monitoring program for California: Assembly Bill 599 Report to the Governor and Legislature, March 2003, 100 p. (Also available at http:// www.waterboards.ca.gov/water_issues/programs/gama/ laws_regs.shtml.)

California State Water Resources Control Board, 2010, GAMA—Groundwater Ambient Monitoring and Assessment Program: State Water Resources Control Board website, accessed September 9, 2010, at http://www. waterboards.ca.gov/gama/.

Chapelle, F.H., 2001, Groundwater microbiology and geochemistry (2d ed.): New York, John Wiley and Sons, Inc., 477 p.

Chapelle, F.H., McMahon, P.B., Dubrovsky, N.M., Fuji, R.F., Oaksford, E.T., and Vroblesky, D.A., 1995, Deducing the distribution of terminal electron-accepting processes in hydrologically diverse groundwater systems: Water Resources Research, v. 31, no. 2, p. 359–371.

Clark, I.D., and Fritz, P., 1997, Environmental isotopes in hydrogeology: New York, Lewis Publishers, 328 p.

Cook, P.G., and Böhlke, J.K., 2000, Determining timescales for groundwater flow and solute transport, in Cook, P.G., and Herczeg, A., eds., Environmental tracers in subsurface hydrology: Boston, Kluwer Academic Publishers, p. 1–30.

Craig, H., and Lal, D., 1961, The production rate of natural tritium: Tellus, v. 13, p. 85–105.

Davis, S., and DeWiest, R.J., 1966, Hydrogeology: New York, John Wiley and Sons, 413 p.

Domagalski, J.L, and Dubrovsky, N.M., 1991, Regional assessment of non point-source pesticide residues in ground water, San Joaquin Valley, California: U.S. Geological Survey Water-Resources Investigations Report 91–4027, 14 p.

Fontes, J.C., and Garnier, J.M., 1979, Determination of the initial 14C activity of the total dissolved carbon—A review of the existing models and a new approach: Water Resources Research, v. 15, p. 399–413.

Fram, M.S., and Belitz, Kenneth, 2011, Occurrence and concentrations of pharmaceutical compounds in groundwater used for public drinking-water supply in California: Science of the Total Environment, v. 409, no. 18, p. 3409–3417. Available at: http://www.sciencedirect.com/science/article/pii/S0048969711005778.

Gilliom, R.J., Barbash, J.E., Crawford, C.G., Hamilton, P.A., Martin, J.D., Nakagaki, N., Nowell, L.H., Scott, J.C., Stackelberg, P.E., Thelin, G.P., and Wolock, D.M., 2006, The quality of our nation's waters—Pesticides in the nation's streams and ground water, 1992–2001: U.S. Geological Survey Circular 1291, 172 p.

Hem, J.D., 1970, Study and interpretation of the chemical characteristics of natural water, (2d ed.): U.S. Geological Survey Water-Supply Paper 1473, 363 p.

Isaaks, E.H., and Srivastava, R.M., 1989, Applied geostatistics: New York, Oxford University Press, 511 p.

Johnson, T.D., and Belitz, Kenneth, 2009, Assigning land use to supply wells for the statistical characterization of regional groundwater quality—Correlating urban land use and VOC occurrence: Journal of Hydrology, v. 370, p. 100–108.

Jurgens, B.C., McMahon, P.B., Chapelle, F.H., and Eberts, S.M., 2009, An Excel® workbook for identifying redox processes in ground water: U.S. Geological Survey Open-File Report 2009–1004, 8 p. (Also available at http://pubs.usgs.gov/of/2009/1004/.)

Kulongoski, J.T., and Belitz, Kenneth, 2004, Ground-water ambient monitoring and assessment program: U.S. Geological Survey Fact Sheet 2004–3088, 4 p. (Also available at http://pubs.usgs.gov/fs/2004/3088/.)

Kulongoski, J.T., and Belitz, Kenneth, 2011, Status and understanding of groundwater quality in the Monterey Bay and Salinas Valley Basins, 2005—California GAMA Priority Basin Project: U.S. Geological Survey Scientific Investigations Report 2011–5058, 84 p. (Also available at http://pubs.usgs.gov/sir/2011/5058/.)

Kulongoski, J.T., Belitz, Kenneth, Landon, M.K., and Farrar, Christopher, 2010, Status and understanding of groundwater quality in the North San Francisco Bay groundwater basins, 2004—California GAMA Priority Basin Project: U.S. Geological Survey Scientific Investigations Report 2010–5089, 87 p. (Also available at http://pubs.usgs.gov/sir/2010/5089/.)

Kulongoski, J.T., Hilton, D.R., Cresswell, R.G., Hostetler, S., and Jacobson, G., 2008, Helium-4 characteristics of groundwaters from Central Australia—Comparative chronology with chlorine-36 and carbon-14 dating techniques: Journal of Hydrology, v. 348, p. 176–194.

Land, Michael, and Belitz, Kenneth, 2008, Ground-water quality data in the San Fernando–San Gabriel study unit, 2005—Results from the California GAMA Program: U.S. Geological Survey Data Series 356, 84 p. (Also available at http://pubs.usgs.gov/ds/356/.)

Landon, M.K., Belitz, Kenneth, Jurgens, B.C., Kulongoski, J.T., and Johnson, T.D., 2010, Status and understanding of groundwater quality in the Central–Eastside San Joaquin basin, 2006—California GAMA Priority Basin Project: U.S. Geological Survey Scientific Investigations Report 2009–5266, 97 p. (Also available at http://pubs.usgs.gov/sir/2009/5266/.)

Lindberg, R.D., and Runnells, D.D., 1984, Ground water redox reactions—An analysis of equilibrium state applied to Eh measurements and geochemical modeling: Science, v. 225, p. 925–927.

Lucas, L.L., and Unterweger, M.P., 2000, Comprehensive review and critical evaluation of the half-life of tritium: Journal of Research of the National Institute of Standards and Technology, v. 105, no. 4, p. 541–549.

Manning, A.H., Solomon, D.K., and Thiros, S.A., 2005, 3H/3He age data in assessing the susceptibility of wells to contamination: Ground Water, v. 43, no. 3, p. 353–367.

McMahon, P.B., and Chapelle, F.H., 2008, Redox processes and water quality of selected principal aquifer systems: Ground Water, v. 46, no. 2, p. 259–271.

Michel, R.L., 1989, Tritium deposition in the continental United States, 1953–83: U.S. Geological Survey Water-Resources Investigations Report 89–4072, 46 p.

Michel, R.L., and Schroeder, R., 1994, Use of long-term tritium records from the Colorado River to determine timescales for hydrologic processes associated with irrigation in the Imperial Valley, California: Applied Geochemistry, v. 9, p. 387–401.

Morrison, P., and Pine, J., 1955, Radiogenic origin of the helium isotopes in rock: Annals of the New York Academy of Sciences, v. 12, p. 19–92.

Nakagaki, Naomi, Price, C.V., Falcone, J.A., Hitt, K.J., and Ruddy, B.C., 2007, Enhanced National Land Cover Data 1992 (NLCDe 92): U.S. Geological Survey raster digital data, available online at http://water.usgs.gov/lookup/getspatial?nlcde92.

Nakagaki, Naomi, and Wolock, D.M., 2005, Estimation of agricultural pesticide use in drainage basins using land cover maps and county pesticide data: U.S. Geological Survey Open-File Report 2005–1188, 56 p. (Also available at http://pubs.usgs.gov/of/2005/1188/.)

Piper, A.M., 1944, A graphic procedure in the geochemical interpretation of water analyses: American Geophysical Union Transactions, v. 25, p. 914–923.

Plummer, L.N., Michel, R.L., Thurman, E.M., and Glynn, P.D., 1993, Environmental tracers for age-dating young ground water, in Alley, W.M., ed., Regional groundwater quality: New York, Van Nostrand Reinhold, p. 255–294.

Poreda, R.J., Cerling, T.E., and Salomon, D.K., 1988, Tritium and helium isotopes as hydrologic tracers in a shallow unconfined aquifer: Journal of Hydrology, v. 103, p. 1–9.

Rowe, B.L., Toccalino, P.L., Moran, M.J., Zogorski, J.S., and Price, C.V., 2007, Occurrence and potential human-health relevance of volatile organic compounds in drinking water from domestic wells in the United States: Environmental Health Perspectives, v. 115, no. 11, p. 1539–1546.

Schlosser, P., Stute, M., Dorr, H., Sonntag, C., and Oto, K.M., 1988, Tritium/3He dating of shallow groundwater: Earth Planetary Science Letters, v. 89, p. 353–362.

Scott, J.C., 1990, Computerized stratified random site selection approaches for design of a ground-water-quality sampling network: U.S. Geological Survey Water-Resources Investigations Report 90–4101, 109 p.

State of California, 1999, Supplemental Report of the 1999 Budget Act 1999-00 Fiscal Year, Item 3940-001-0001, State Water Resources Control Board, accessed September 9, 2010, at http://www.lao.ca.gov/1999/99-00_supp_rpt_lang.html#3940.

State of California, 2001a, Assembly Bill No. 599, Chapter 522, accessed September 9, 2010, at http://www.swrcb.ca.gov/gama/docs/ab_599_bill_20011005_chaptered.pdf.

State of California, 2001b, Groundwater Monitoring Act of 2001: California Water Code, part 2.76, Sections 10780-10782.3, accessed September 9, 2010, at http://www.leginfo.ca.gov/cgi-bin/displaycode?section=wat&group=10001-11000&file=10780-10782.3.

Takaoka, N., and Mizutani, Y., 1987, Tritiogenic 3He in groundwater in Takaoka: Earth and Planetary Science Letters, v. 85, p. 74–78.

Tinsley, J.C., III, 2001, Aspects of the quaternary geology of the San Fernando Valley, California [abs.], in Proceedings of the American Association of Petroleum Geologists Pacific Section Meeting, Universal City, California, April 9–11, 2001. (Also available at http://www.searchanddiscovery.com/abstracts/html/2001/pacific/abstracts/tinsley.htm.)

Toccalino, P.L., and Norman, J.E., 2006, Health-based screening levels to evaluate U.S. Geological Survey ground-water quality data: Risk Analysis, v. 26, no. 5, p. 1339–1348.

Toccalino, P.L., Norman, J.E., Phillips, R.H., Kauffman, L.J., Stackelberg, P.E., Nowell, L.H., Krietzman, S.J., and Post, G.B., 2004, Application of health-based screening levels to ground-water quality data in a state-scale pilot effort: U. S. Geological Survey Scientific Investigations Report 2004–5174, 64 p. (Also available at http://pubs.usgs.gov/sir/2004/5174/.)

Tolstikhin, I.N., and Kamensky, I.L., 1969, Determination of groundwater ages by the T-3He method: Geochemistry International, v. 6, p. 310–811.

Torgersen, T., 1980, Controls on pore-fluid concentrations of 4He and 222Rn and the calculation of 4He/222Rn ages: Journal of Geochemical Exploration, v. 13, p. 7–75.

Torgersen, T., and Clarke, W.B., 1985, Helium accumulation in groundwater—I. An evaluation of sources and continental flux of crustal ^{4}He in the Great Artesian basin, Australia: Geochimica et Cosmochimica Acta, v. 49, p. 1211–1218.

Torgersen, T., Clarke, W.B., and Jenkins, W.J.,1979, The tritium/helium-3 method in hydrology, in Isotope Hydrology 1978: IAEA-SM-228/2, IAEA, Vienna, p. 917–930.

Troiano, J., Weaver, D., Marade, J., Spurlock, F., Pepple, M., Nordmark, C., and Bartkowiak, D., 2001, Summary of well water sampling in California to detect pesticide residues resulting from nonpoint source applications: Journal of Environmental Quality, v. 30, p. 448–459.

U.S. Environmental Protection Agency, 1998, Reporting requirements for risk/benefit information: Federal Register, v. 62, no. 182, p. 49369–49395, accessed July 7, 2009, at http://www.epa.gov/EPA-PEST/1997/September/Day-19/p24937.htm.

U.S. Environmental Protection Agency, 2006, 2006 Edition of the drinking water standards and health advisories: Washington, D.C., U.S. Environmental Protection Agency, Office of Water EPA/822/R-06–013. (Also available at http://www.epa.gov/waterscience/criteria/drinking/dwstandards.pdf.)

U.S. Geological Survey, 1996, USGS response to an urban earthquake—Northridge '94: U.S. Geological Survey Open-File Report 96-263. (Available at http://pubs.usgs.gov/of/1996/ofr-96-0263/.)

U.S. Geological Survey, 2010, What is the Priority Basin Project?: U.S. Geological Survey website, accessed September 9, 2010, at http://ca.water.usgs.gov/gama.

Vogel, J.C., and Ehhalt, D., 1963, The use of the carbon isotopes in groundwater studies—Radioisotopes in Hydrology: Vienna, IAEA, p. 383–395.

Zogorski, J.S., Carter, J.M., Ivahnenko, Tamara, Lapham, W.W., Moran, M.J., Rowe, B.L., Squillace, P.J., and Toccalino, P.L., 2006, The quality of our Nation's waters—Volatile organic compounds in the Nation's ground water and drinking-water supply wells: U.S. Geological Survey Circular 1292, 101 p.

Appendix A. Ancillary Datasets

Land-Use Classification

Land use was classified using an enhanced version of the satellite-derived [98 ft (30 m) pixel resolution] USGS National Land Cover Dataset (Nakagaki and others, 2007). This dataset has been used in previous national and regional studies relating land use to water quality (Gilliom and others, 2006; Zogorski and others, 2006). The dataset characterizes land cover during the early 1990s. One pixel in the dataset imagery represents a land area of 9,688 ft^2 (900 m^2), calculated from the pixel size of 98 ft (30 m). The imagery was classified into 25 land-cover classifications (Nakagaki and Wolock, 2005). These 25 land-cover classifications were aggregated into four principal land-use classes—urban, agricultural, natural, and mixed. Each pixel was assigned a land-use class if greater than 50 percent of the land cover in that area could be associated with a single land use. If no land cover was greater than 50 percent of the pixel area, the classification of "mixed" was assigned.

Land-use classes for the study unit, for study areas, and for areas within a radius of 1,640 ft (500 m) surrounding each well were assigned using the USGS National Land Cover Dataset (Johnson and Belitz, 2009). Land-use classes for the study unit and the study areas (fig. 5) were calculated from the land cover of each pixel in the study unit and the study areas. Land use assigned to the area surrounding an individual well (table A1) was calculated from land use within the area surrounding each well [radius of 1,640 ft (500 m) and land area of 8,449,620 ft^2 (785,400 m^2)].

Well-Construction Information

Identification numbers of wells where groundwater samples were analyzed for the FG study unit are listed in table A2. Available well-construction information for wells is presented in table A3. Other sources of well-construction information were ancillary records from well owners, the USGS National Water Information System database, and the CDPH database. Well selection procedures are described by Land and Belitz (2008).

Groundwater-Age Classification

Groundwater recharge temperatures from noble gases, age data, and classifications are listed in table A4. Groundwater dating techniques indicate the time since the groundwater was last in contact with the atmosphere. Techniques used to estimate groundwater residence times or 'age' include those based on tritium (for example, Tolstikhin and Kamensky, 1969), tritium combined with its decay product helium-3 (for example, Takaoka and Mizutani,

1987; Poreda and others, 1988, Schlosser and others, 1988), carbon-14 activities (for example, Vogel and Ehhalt, 1963; Plummer and others, 1993), and dissolved noble gases, particularly helium-4 accumulation (for example, Davis and DeWiest, 1966; Andrews and Lee, 1979; Kulongoski and others, 2008).

Tritium (^{3}H) is a short-lived radioactive isotope of hydrogen with a half-life of 12.32 years (Lucas and Unterweger, 2000). Tritium is produced naturally in the atmosphere from the interaction of cosmogenic radiation with nitrogen (Craig and Lal, 1961), by above-ground nuclear explosions, and by the operation of nuclear reactors. Tritium enters the hydrological cycle following oxidation to tritiated water (HTO). Consequently, the presence of ^{3}H in groundwater may be used to identify water that has been exposed to the atmosphere since 1952. By determining the ratio of ^{3}H to ^{3}He, resulting from the radioactive decay of ^{3}H, the time that the water has resided in the aquifer can be calculated more precisely than by using tritium alone (for example, Takaoka and Mizutani, 1987; Poreda and others, 1988).

Carbon-14 (^{14}C) is a widely used chronometer based on the radiocarbon content of dissolved inorganic carbonate species in groundwater. ^{14}C is formed in the atmosphere by the interaction of cosmic-ray neutrons with nitrogen and, to a lesser degree, with oxygen and carbon. ^{14}C is incorporated into carbon dioxide and mixed throughout the atmosphere, dissolved in precipitation, and incorporated into the hydrologic cycle. ^{14}C activity in groundwater, expressed as percent modern carbon (pmc), reflects exposure to the atmospheric ^{14}C source and is governed by the decay constant of ^{14}C (with a half-life of 5,730 years). ^{14}C can be used to estimate groundwater ages ranging from 1,000 to less than 30,000 years before present because of its half-life. Calculated ^{14}C ages in this study are referred to as "uncorrected" because they have not been adjusted to consider exchanges with sedimentary sources of carbon (Fontes and Garnier, 1979).The ^{14}C age (residence time) is calculated on the basis of the decrease in ^{14}C activity as a result of radioactive decay since groundwater recharge, relative to an assumed initial ^{14}C concentration (Clarke and Fritz, 1997). A mean initial ^{14}C activity of 99 pmc was assumed for this study, with estimated errors on calculated groundwater ages of as much as ±20 percent.

Helium (He) is a naturally occurring inert gas initially included during the accretion of the planet, and later produced by the radioactive decay of lithium, thorium, and uranium in the Earth. Measured He concentrations in groundwater are the sum of several He components including air-equilibrated He (He$_{eq}$), He from dissolved-air bubbles (He$_a$), terrigenic He (He$_{ter}$), and tritiogenic He-3 (^{3}He$_t$). Helium (^{3}He and ^{4}He) concentrations in groundwater often exceed the expected solubility equilibrium values, which are functions of the temperature of the water, as a result of subsurface production of both isotopes and their subsequent release into the groundwater (for example, Morrison and Pine, 1955; Andrews and Lee, 1979; Torgersen, 1980; Andrews, 1985; Torgersen and Clarke, 1985). The presence of terrigenic He in groundwater, from its production in aquifer material or deeper in the crust, is indicative of long groundwater residence times. The amount of terrigenic helium is defined as the concentration of the total measured helium minus the fraction as a result of air-equilibration [He_{eq}] and dissolved air-bubbles [He_a]. For the purposes of this study, percentage of terrigenic He is used to identify groundwater with residence times greater than 100 years. Percentage of terrigenic He is defined as the concentration of terrigenic He (as defined previously) divided by the total measured He in the sample (corrected for air-bubble entrainment). Samples with greater than 5 percent terrigenic He indicate that groundwater has a residence time of more than 100 years.

Recharge temperatures for 24 samples were calculated from dissolved neon, argon, krypton, and xenon data by using methods described by Aeschbach-Hertig and others (1999). The only modeled recharge temperatures accepted were those for which the probability was greater than 1 percent that the sum of the squared deviations between the modeled and the measured concentrations (weighted with the experimental 1-sigma errors) was equal to or greater than the observed value (Aeschbach-Hertig and others, 2000). The recharge temperature with the highest probability for each sample was used in this report.

$^3H/^3He$ ages were computed as described by Poreda and others (1988). The $^3He/^4He$ of samples was determined by the linear regression of the percentage of terrigenic He and δ^3He [($\delta 3He = R_{meas}/R_{atm} -1$) x 100 percent] of samples containing less than 1 tritium unit. Calculations of the noble gas temperature and ^3He to ^4He ratios are useful because they constrain helium-based groundwater ages further.

In this study, the ages of samples are classified as pre-modern, modern, and mixed. Groundwater with tritium activity less than 1 tritium unit (TU), percentage of terrigenic He greater than 5 percent, and ^{14}C less than 90 pmc was designated as pre-modern, defined as having been recharged before 1952. Groundwater with tritium activities greater than 1 TU, percentage of terrigenic He less than 5 percent, and ^{14}C greater than 90 pmc is designated as modern, defined as having been recharged after 1952. Samples with pre-modern and modern components are designated as mixed groundwater, which includes substantial fractions of old and young waters. In reality, pre-modern groundwater could contain very small fractions of modern water and modern water could contain small fractions of pre-modern water. Previous investigations have used a range of tritium values from 0.3 to 1.0 TU as thresholds for distinguishing pre-1952 from post-1952 water (Michel, 1989; Plummer and others, 1993; Michel and Schroeder, 1994; Clark and Fritz, 1997; Manning and others, 2005). By using a tritium value of 1.0 TU for the threshold in this study, the age classification scheme allows a slightly

larger fraction of modern water to be classified as pre-modern than if a lower threshold were used. A lower threshold for tritium would result in fewer samples classified as pre-modern, than mixed, when other tracers, such as ^{14}C and terrigenic helium, would suggest that they were primarily pre-modern. This higher threshold was considered more appropriate for this study because many of the wells were production wells with long screens and mixing of waters of different ages is likely to occur.

Tritium concentrations, percentage of modern carbon, percentage of terrigenic helium, and sample age classifications are reported in table A4. Because of uncertainties in age distributions, in particular those caused by mixing waters of different ages in wells with long perforation intervals and high withdrawal rates, these age estimates were not specifically used for statistically quantifying the relation between age and water quality in this report. Although more sophisticated lumped parameter models used for analyzing age distributions that incorporate mixing are available (for example, Cook and Böhlke, 2000), use of these alternative models to characterize age mixtures was beyond the scope of this report. Rather, classification into modern (recharged after 1952), mixed, and pre-modern (recharged before 1952) categories was sufficient to provide an appropriate and useful characterization for the purposes of examining groundwater quality.

Classification of Geochemical Condition

Geochemical conditions investigated as potential explanatory variables in this report include oxidation-reduction characteristics, and dissolved oxygen (table A5). An automated workbook program was used to assign the redox classification to each sample (Jurgens and others, 2009). Oxidation-reduction (redox) conditions influence the mobility of many organic and inorganic constituents (McMahon and Chapelle, 2008). Along groundwater flow paths, redox conditions commonly proceed along a well-documented sequence of terminal electron acceptor processes (TEAP); one TEAP typically dominates at a particular time and aquifer location (Chapelle and others, 1995; Chapelle, 2001). The predominant TEAPs are oxygen-reduction, nitrate-reduction, manganese-reduction, iron-reduction, sulfate-reduction, and methanogenesis. The presence of redox-sensitive chemical species suggesting more than one TEAP may indicate mixed waters from different redox zones upgradient of the well, a well screened across more than one redox zone, or spatial heterogeneity in microbial activity in the aquifer. Different redox elements (for example, iron, manganese, and sulfur) tend not to reach overall equilibrium in most natural water systems (Lindberg and Runnells, 1984); therefore, a single redox measurement usually cannot represent the system, further complicating the assessment of redox conditions. pH is the measure of hydrogen-ion activity in a water sample and is sensitive to redox conditions.

Table A1.   Percent land use by category, land-use classification, cell number, and USGS GAMA well identification number for well data used in the San Fernando–San Gabriel study unit, California GAMA Priority Basin Project.

[SF, San Fernando study area well; SG, San Gabriel study area well; SFU or SGU, understanding well; USGS, U.S. Geological Survey; a USGS GAMA well identification number indicates the use of USGS data from the grid well; and a well identification number with 'DPH' indicates the use of California Department of Public Health (CDPH) data from a CDPH well in the cell]

Grid cell number	USGS-GAMA well identification number indicating data source	Land-use catagories (within 500 m of the well, in percent)			Land-use classification
		Agricultural	Natural	Urban	
San Fernando study area grid wells					
4	SF-02		7	93	Urban
5	SF-03		5	95	Urban
6	SF-04		32	68	Urban
7	SF-05		3	97	Urban
8	SF-06		1	99	Urban
9	SF-10		1	99	Urban
10	SF-07		9	91	Urban
11	SF-01		96	4	Natural
11	SF-DPH-13		95	5	Natural
12	SF-11		99	1	Natural
13	SF-09		9	91	Urban
14	SF-08		18	82	Urban
15	SF-12		11	89	Urban
San Gabriel study area grid wells					
1	SG-13		2	98	Urban
2	SG-10		6	94	Urban
2	SG-DPH-24	1	14	85	Urban
3	SG-08		3	97	Urban
4	SG-11			100	Urban
5	SG-07		16	84	Urban
5	SG-DPH-25		12	88	Urban
6	SG-18		28	72	Urban
6	SG-DPH-26		1	99	Urban
7	SG-22		8	92	Urban
8	SG-12		15	85	Urban
9	SG-19		10	90	Urban
10	SG-06	17	47	36	Mixed
12	SG-05		18	82	Urban
12	SG-DPH-27	1	32	66	Urban
13	SG-04		1	99	Urban
14	SG-09		15	85	Urban
14	SG-DPH-28		23	77	Urban
15	SG-14		49	51	Urban
16	SG-15	1	52	47	Natural
17	SG-21		2	98	Urban
18	SG-20	1	15	84	Urban
19	SG-01		13	87	Urban
21	SG-17	32	18	50	Mixed
22	SG-03	1	16	83	Urban
22	SG-DPH-29	27	22	51	Urban
23	SG-16		7	93	Urban
24	SG-23		37	63	Urban
25	SG-02		25	75	Urban
25	SG-DPH-30		29	71	Urban

Table A1. Percent land use by category, land-use classification, cell number, and USGS GAMA well identification number for well data used in the San Fernando–San Gabriel study unit, California GAMA Priority Basin Project.—Continued

[SF, San Fernando study area well; SG, San Gabriel study area well; SFU or SGU, understanding well; USGS, U.S. Geological Survey; a USGS GAMA well identification number indicates the use of USGS data from the grid well; and a well identification number with 'DPH' indicates the use of California Department of Public Health (CDPH) data from a CDPH well in the cell]

Grid cell number	USGS-GAMA well identification number indicating data source	Land-use catagories (within 500 m of the well, in percent)			Land-use classification
		Agricultural	Natural	Urban	
USGS understanding wells					
6	SFU-04		21	79	Urban
7	SFU-06		1	99	Urban
10	SFU-05		42	58	Urban
13	SFU-01		10	90	Urban
13	SFU-02	1	38	62	Urban
15	SFU-03		1	99	Urban
3	SGU-07		2	98	Urban
4	SGU-06			100	Urban
8	SGU-05		4	96	Urban
8	SGU-08			100	Urban
10	SGU-04	1	21	79	Urban
13	SGU-09		15	85	Urban
14	SGU-11		11	89	Urban
16	SGU-03		31	68	Urban
23	SGU-01		5	95	Urban
25	SGU-02		29	71	Urban
na	SGU-10		11	89	Urban

Table A2.   Cell number and USGS-GAMA well identification numbers for well data used in the San Fernando–San Gabriel study unit, California GAMA Priority Basin Project.

[A USGS-GAMA well identification number indicates the use of USGS data from the grid well; a CDPH-GAMA well identification number with 'DG' signifies the use of CDPH inorganic data from the grid well; a CDPH-GAMA well identification number with 'DPH' indicates the use of CDPH data from a different well. SF, San Fernando study area well; SG, San Gabriel study area well; SFU or SGU, understanding well; --, no wells sampled or selected; USGS, U.S. Geological Survey]

Grid cell number	USGS-GAMA well identification number	Grid supplemented by CDPH data from a USGS-grid well	Grid supplemented by CDPH data from different well	Grid cell number	USGS-GAMA well identification number	Grid supplemented by CDPH data from a USGS-grid well	Grid supplemented by CDPH data from different well
San Fernando study area grid wells				**San Gabriel study area grid wells—Continued**			
1	–	–	–	15	SG-14	–	–
2	–	–	–	16	SG-15	–	–
3	–	–	–	17	SG-21	SG-DG-21	–
4	SF-02	SF-DG-02	–	18	SG-20	SG-DG-20	–
5	SF-03	SF-DG-03	–	19	SG-01	SG-DG-01	–
6	SF-04	SF-DG-04	–	20	–	–	–
7	SF-05	SF-DG-05	–	21	SG-17	SG-DG-17	–
8	SF-06	SF-DG-06	–	22	SG-03	–	SG-DPH-29
9	SF-10	–	–	23	SG-16	–	–
10	SF-07	SF-DG-07	–	24	SG-23	SG-DG-23	–
11	SF-01	–	SF-DPH-13	25	SG-02	–	SG-DPH-30
12	SF-11	–	–	**USGS understanding wells**			
13	SF-09	–	–	6	SFU-04	–	–
14	SF-08	–	–	7	SFU-06	–	–
15	SF-12	SF-DG-12	–	10	SFU-05	–	–
San Gabriel study area grid wells				13	SFU-01	–	–
1	SG-13	SG-DG-13	–	13	SFU-02	–	–
2	SG-10	–	SG-DPH-24	15	SFU-03	–	–
3	SG-08	–	–	3	SGU-07	–	–
4	SG-11	SG-DG-11	–	4	SGU-06	–	–
5	SG-07	–	SG-DPH-25	8	SGU-05	–	–
6	SG-18	–	SG-DPH-26	8	SGU-08	–	–
7	SG-22	SG-DG-22	–	10	SGU-04	–	–
8	SG-12	SG-DG-12	–	13	SGU-09	–	–
9	SG-19	SG-DG-19	–	14	SGU-11	–	–
10	SG-06	–	–	16	SGU-03	–	–
11	–	–	–	23	SGU-01	–	–
12	SG-05	–	SG-DPH-27	25	SGU-02	–	–
13	SG-04	SG-DG-04	–	10	SGU-10	–	–
14	SG-09	–	SG-DPH-28				

Table A3.   Well construction information for wells used in the San Fernando–San Gabriel study unit, California GAMA Priority Basin Project.

[SF, San Fernando study area well; SG, San Gabriel study area well; SFU or SGU, understanding well; USGS, U.S. Geological Survey; PSW, public supply well; IRR, irrigation well; IND, industrial well; na, data not available]

Grid cell number	USGS-GAMA well identification number indicating data source	Construction information (in feet below LSD)			
		Well depth	Top of perforations	Bottom of perforations	Length from top of uppermost perforated interval to bottom perforation
San Fernando study area grid wells					
4	SF-02	488	230	435	205
5	SF-03	377	250	355	105
6	SF-04	800	400	780	380
7	SF-05	594	195	578	383
8	SF-06	490	242	418	176
9	SF-10	930	268	894	626
10	SF-07	359	109	349	240
11	SF-01	600	185	486	301
11	SF-DPH-13	na	na	na	na
12	SF-11	480	80	480	400
13	SF-09	184	50	170	120
14	SF-08	400	200	380	180
15	SF-12	199	84	174	90
San Gabriel study area grid wells					
1	SG-13	970	320	970	650
2	SG-10	785	291	762	471
2	SG-DPH-24	na	na	na	na
3	SG-08	399	110	299	189
4	SG-11	490	160	365	205
5	SG-07	587	260	587	327
5	SG-DPH-25	na	na	na	na
6	SG-18	700	286	585	299
6	SG-DPH-26	600	260	580	320
7	SG-22	785	380	765	385
8	SG-12	600	229	600	371
9	SG-19	804	260	804	544
10	SG-06	712	166	712	546
12	SG-05	1,290	1,013	1,275	262
12	SG-DPH-27	na	na	na	na
13	SG-04	198	128	193	65
14	SG-09	600	300	580	280
14	SG-DPH-28	540	73	420	347
15	SG-14	1,000	500	1,000	500
16	SG-15	400	115	340	225
17	SG-21	1,152	792	1,132	340
18	SG-20	500	198	484	286
19	SG-01	810	670	790	120
21	SG-17	186	na	na	na
22	SG-03	372	62	370	308
22	SG-DPH-29	na	na	na	na

Table A3. Well construction information for wells used in the San Fernando–San Gabriel study unit, California GAMA Priority Basin Project.—Continued

[SF, San Fernando study area well; SG, San Gabriel study area well; SFU or SGU, understanding well; USGS, U.S. Geological Survey; PSW, public supply well; IRR, irrigation well; IND, industrial well; na, data not available]

Grid cell number	USGS-GAMA well identification number indicating data source	Construction information (in feet below LSD)			
		Well depth	Top of perforations	Bottom of perforations	Length from top of uppermost perforated interval to bottom perforation
San Gabriel study area grid wells—Continued					
23	SG-16	414	157	233	76
24	SG-23	300	na	na	na
25	SG-02	480	358	480	122
25	SG-DPH-30	na	na	na	na
USGS Understanding wells					
6	SFU-04	800	400	780	380
7	SFU-06	800	370	770	400
10	SFU-05	610	310	600	290
13	SFU-01	267	138	248	110
13	SFU-02	196	110	196	86
15	SFU-03	268	100	253	153
3	SGU-07	800	450	780	330
4	SGU-06	1,089	270	1,058	788
8	SGU-05	1,008	360	1,008	648
8	SGU-08	942	292	918	626
10	SGU-04	664	178	400	222
13	SGU-09	600	298	581	283
14	SGU-11	580	250	580	330
16	SGU-03	507	280	500	220
23	SGU-01	699	143	437	294
25	SGU-02	312	40	312	272
na	SGU-10	264	25	260	235

Table A4. Noble-gas-based recharge temperature, tritium, terrigenic helium, percent modern carbon, and age classification of samples, San Fernando–San Gabriel study unit, California GAMA Priority Basin Project.

[modern, recharged after 1952; mixed, modern and pre-modern water; pre-modern, recharged prior to 1952; °C, degrees Celsius; nc, not collected; <, less than]

Well identification number	Noble-gas-based recharge temperature, °C	Tritium, tritium units	Terrigenic helium, percent of total helium	Percent modern carbon	Age classification
SF-08	17.6	1.0	61.9	78.4	Mixed
SF-09	18.4	6.4	3.1	99.5	Modern
SF-10	15.1	1.2	83.2	79.1	Mixed
SF-12	19.9	2.2	17.9	nc	Mixed
SFU-03	19.1	5.0	22.0	104.0	Mixed
SFU-04	16.9	3.4	0.0	104.3	Modern
SFU-06	17.2	2.10	0.0	101.7	Modern
SG-05	16.4	0.1	38.8	nc	Pre-modern
SG-06	19.5	6.1	0.0	91.9	Modern
SG-07	14.6	5.0	0.0	nc	Modern
SG-08	18.1	1.9	0.0	93.0	Modern
SG-14	15.1	4.2	0.0	97.1	Modern
SG-15	13.5	4.6	0.0	96.5	Modern
SG-16	22.1	3.8	0.0	103.1	Modern
SGU-01	13.8	2.4	1.8	nc	Modern
SGU-02	16.2	4.4	30.8	104.6	Mixed
SGU-04	17.7	0.9	51.8	83.2	Pre-modern
SGU-05	17.8	0.0	20.7	85.8	Pre-modern
SGU-06	15.8	2.8	67.5	nc	Mixed
SGU-07	16.3	0.4	86.4	72.4	Pre-modern
SGU-08	15.8	0.1	15.8	nc	Pre-modern
SGU-09	15.2	4.2	0.0	98.9	Modern
SGU-10	19.0	7.1	0.0	nc	Modern
SGU-11	14.2	5.3	0.0	99.9	Modern

Table A5. Oxidation-reduction constituents and redox classification for samples from the San Fernando–San Gabriel study unit, California GAMA Priority Basin Project.

[Indeterminant, insufficient data to determine redox classification; mg/L, milligram per liter; μg/L, microgram per liter; na, data not available; oxic, dissolved oxygen ≥ 0.5 mg/L; redox, oxidation-reduction; >, greater than; ≥, greater than or equal to; <, less than; USGS, U.S. Geological Survey; CDPH, California Department of Public Health]

USGS-GAMA well identification number[1]	Oxidation-reduction constituent					Redox classification
	Dissolved oxygen	Nitrate, as nitrogen	Manganese	Iron	Sulfate	
	Oxidation-reduction threshold value					
	≥0.5	>0.5	>50	>100	>4.0	
	Possible redox type if concentration > redox threshold value					
	O_2	NO_3	Mn	Fe	SO_4	
	Analysis reporting level and associated units					
	0.1	0.06	0.18	5.0	0.18	
	mg/L	mg/L	μg/L	μg/L	mg/L	
San Fernando study area grid wells						
SF-DG-02	7.77	5.8	2.1	na	<	53.0	Oxic
SF-DG-03	7.9	7.4	8.6	na	na	[2] 45	Oxic
SF-DG-04	7.58	1.5	2.2	<	<	87.4	Oxic
SF-DG-05	7.64	0.5	1.8	na	<	[2] 468	Oxic
SF-DG-06	na	0.6	2.0	na	na	na	Oxic
SF-10	7.7	4.2	1.9	1.1	<	101	Oxic
SF-DG-07	7.51	4.4	7.4	56.3	1,890	62.9	Mixed
SF-DPH-13	8.26	7.0	2.4	na	na	21.9	Oxic
SF-11	na	1.5	na	na	na	na	Oxic
SF-09	6.6	5.8	9.7	0.1	4	166	Oxic
SF-08	7.4	4.2	3.0	9.7	5	122	Oxic
SF-12	na	6.9	9.0	0.1	<	103	Oxic
San Gabriel study area grid wells						
SG-DG-13	7.5	1.8	0.8	na	na	19.0	Oxic
SG-DPH-24	7.2	8.2	4.7	na	na	19.0	Oxic
SG-08	7.5	7.6	10.3	0.8	<	65.0	Oxic
SG-11	na	11.0	5.0	0.1	5	60.7	Oxic
SG-DPH-25	7.86	9.0	0.5	na	na	134	Oxic
SG-DPH-26	7.79	8.3	8.0	na	na	39.0	Oxic
SG-DG-22	7.74	6.2	2.7	na	na	24.3	Oxic
SG-DG-12	7.4	na	1.1	na	na	29.0	Indeterminate
SG-DG-19	7.54	4.9	0.8	na	na	13.0	Oxic
SG-06	7.4	1.5	1.9	0.3	<	116	Oxic
SG-05	na	8.7	0.7	0.1	<	25.7	Oxic
SG-DG-04	7.32	7.8	3.3	na	na	45.0	Oxic
SG-DPH-28	7.5	8.0	1.4	<	<	37.0	Oxic
SG-14	7.6	4.6	1.1	1.6	6	30.1	Oxic
SG-15	7.4	5.2	0.8	<	<	17.0	Oxic
SG-DG-21	7.7	7.9	13.6	na	na	52.0	Oxic
SG-DG-20	7.7	2.5	3.7	na	na	43.0	Oxic
SG-DG-01	7.7	3.4	7.9	na	na	190	Oxic

Table A5   Oxidation-reduction constituents and redox classification for samples from the San Fernando–San Gabriel study unit, California GAMA Priority Basin.—Continued

[Indeterminant, insufficient data to determine redox classification; mg/L, milligram per liter; mg/L, microgram per liter; na, data not available; oxic, dissolved oxygen ≥ 0.5 mg/L; redox, oxidation-reduction; >, greater than; ≥, greater than or equal to; <, less than; USGS, U.S. Geological Survey; CDPH, California Department of Public Health

USGS-GAMA well identification number[1]	Oxidation-reduction constituent					Redox classification
	Dissolved oxygen	Nitrate, as nitrogen	Manganese	Iron	Sulfate	
	Oxidation-reduction threshold value					
	≥0.5	>0.5	>50	>100	>4.0	
	Possible redox type if concentration > redox threshold value					
	O_2	NO_3	Mn	Fe	SO_4	
	Analysis reporting level and associated units					
	0.1	0.06	0.18	5.0	0.18	
	mg/L	mg/L	µg/L	µg/L	mg/L	
San Gabriel study area grid wells—Continued						
SG-DG-17	7.4	2.6	14.0	na	140	180	Oxic
SG-DPH-29	7.9	8.6	5.4	na	na	61.0	Oxic
SG-16	7	6.5	3.6	2.7	23	54.1	Oxic
SG-DG-23	7.48	9.3	1.2	na	na	29.0	Oxic
SG-DPH-30	7.3	12.0	5.0	na	na	150	Oxic
USGS Understanding wells						
SFU-01	na	6.8	na	na	na	na	Oxic
SFU-02	na	5.6	na	na	na	na[2]	Oxic
SFU-03	6.9	4.7	8.7	0.1	<	121	Oxic
SFU-04	7.3	7.6	3.0	<	<	73.3	Oxic
SFU-05	na	3.4	na	na	na	na	Oxic
SFU-06	7.2	8.1	3.1	0.2	<	91.7	Oxic
SGU-01	7.7	10.7	8.5	0.1	<	52.4	Oxic
SGU-02	6.6	3.3	6.1	0.2	<	108	Oxic
SGU-03	na	5.5	na	na	na	na	Oxic
SGU-04	7.5	7.4	1.5	<	<	41.0	Oxic
SGU-05	7.5	6.5	3.6	0.1	<	17.1	Oxic
SGU-06	7.8	6.2	5.8	2	7	66.8	Oxic
SGU-07	8.2	2.6	2.6	<	<	33.1	Oxic
SGU-08	7.7	6.1	1.2	0.7	3	[2] 13.5	Oxic
SGU-09	7.5	4.8	0.5	0.1	5	23.4	Oxic
SGU-10	7	7.2	2.1	34.8	7	121	Oxic
SGU-11	7.5	8.3	1.0	<	<	18.3	Oxic

[1] Values for wells with CDPH GAMA identification are from CDPH database. Values for wells with no CDPH GAMA identification were measured in samples collected by USGS for GAMA. See table 2 for details.

[2] Hydrogen sulfide odor, an indicator of sulfate-reducing conditions, was detected by USGS for GAMA, but not quantified.

Appendix B. Use of Data from the California Department of Public Health (CDPH) Database

For the FG study unit, the historical CDPH database contains more than 1,400,000 records for more than 700 wells, requiring targeted retrievals to manageably use the data to assess water quality. The following paragraphs summarize the selection process for wells and data from the CDPH database for use in the grid-based status assessment (fig. B1).

The strategy used to select CDPH inorganic data for a single well in each cell where the USGS did not obtain a sample for analysis for inorganic constituents involved prioritizing data from different sources. The first choice was to select CDPH data for the grid well sampled by the USGS (fig. B1) for other constituents, provided the CDPH data met quality-control criteria. Cation/anion balance was used as the quality-control assessment metric. Because water is electrically neutral and must have a balance between positive (cations) and negative (anions) electrically charged dissolved species, the cation/anion balance commonly is used as a quality-assurance criterion for water sample analysis (Hem, 1970). An imbalance equal to or greater than 10 percent may indicate uncertainty in the quality of the data or that data were missing for one or more constituents necessary to achieve balance. The most recent CDPH data from the well were evaluated to determine whether the cation/anion imbalance was less than 10 percent; if so, the CDPH inorganic data for the well were selected for use as grid-well data (USGS-grid well with CDPH inorganic data). It was assumed that if analyses met quality-control criteria—cation/anion balance— for major and minor elements, then analyses at these wells for trace elements, nutrients, and radiochemical constituents also would be of acceptable quality. This approach resulted in the selection of inorganic data from the CDPH database for 18 USGS-grid wells. To identify the USGS-grid wells that incorporated CDPH inorganic data, a well ID was created that added "DG" to the GAMA ID for these wells (for example, SF-01 with CDPH data was assigned the well identification SF-DG-01; table A2).

If the first step did not yield CDPH inorganic data for the USGS-grid well, the second step was to search the CDPH database to identify the highest ranked well with a cation/anion imbalance less than 10 percent in each grid cell. This step resulted in selecting CDPH inorganic data for non-USGS-sampled wells for eight grid cells. These eight CDPH wells were not co-located with their cell's respective USGS-grid well. To identify these new CDPH grid wells, a well ID was created that added "DPH" after the study unit prefix and then a sequential number starting after the last GAMA ID for the study area (for example, CDPH well SF-DPH-13, table A1).

If no wells in a grid cell met the cation/anion balance criteria or if there was insufficient data to evaluate charge balance, the third choice for the CDPH well was to select the highest randomly ranked CDPH well with any of the needed inorganic data. This resulted in selecting CDPH inorganic data for three USGS-grid wells. If the well was a USGS-grid well, then a well ID was created that added "DG" to the GAMA ID (for example, SF-DG-06). If the well was a new CDPH well, then "DPH" was added after the study unit prefix and prior to a sequential number starting after the last GAMA ID for the study area (for example, CDPH well SF-DPH-24). In some cases, to achieve one value for each constituent per cell, it was necessary to select an additional well in a cell for data; hence, some cells have multiple CDPH wells.

The result of these steps was one grid well per cell having data from the USGS database, the CDPH database, or both databases. Inorganic data from the CDPH database were used for 23 grid wells. Data were available for 34 grid wells for nitrate plus nitrite and for 7 to 34 wells for most other inorganic constituents (table 2). In combination with USGS-grid well inorganic data (11 wells), inorganic data was available for 34 of the 40 grid cells. Estimates of aquifer-scale proportion for constituents based on a smaller number of wells are subject to a larger error associated with the 90 percent confidence intervals (on the basis of Jeffreys interval for the binomial distribution).

Differences in constituent reporting levels associated with USGS and CDPH data did not affect analysis of high or moderate relative-concentrations because concentrations greater than one-half of water-quality benchmarks were substantially higher than the reporting levels. Several types of comparisons between USGS-collected and CDPH data are described in appendix D.

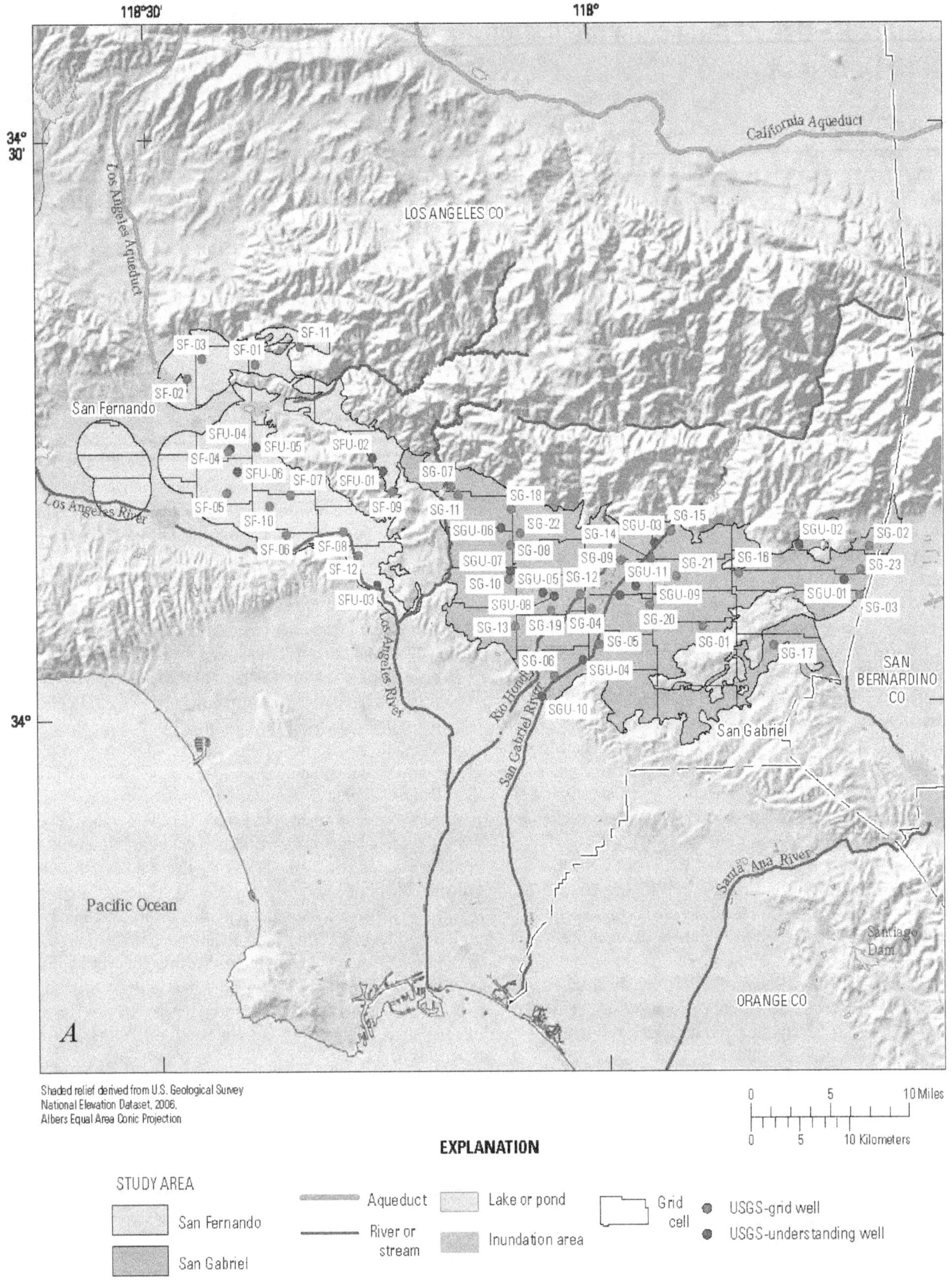

Figure B1.   Map showing identifiers and locations of (*A*) U.S. Geological Survey (USGS) grid and understanding wells, and (*B*) CDPH wells sampled during May–August 2005, San Fernando–San Gabriel study unit, California GAMA Priority Basin Project.

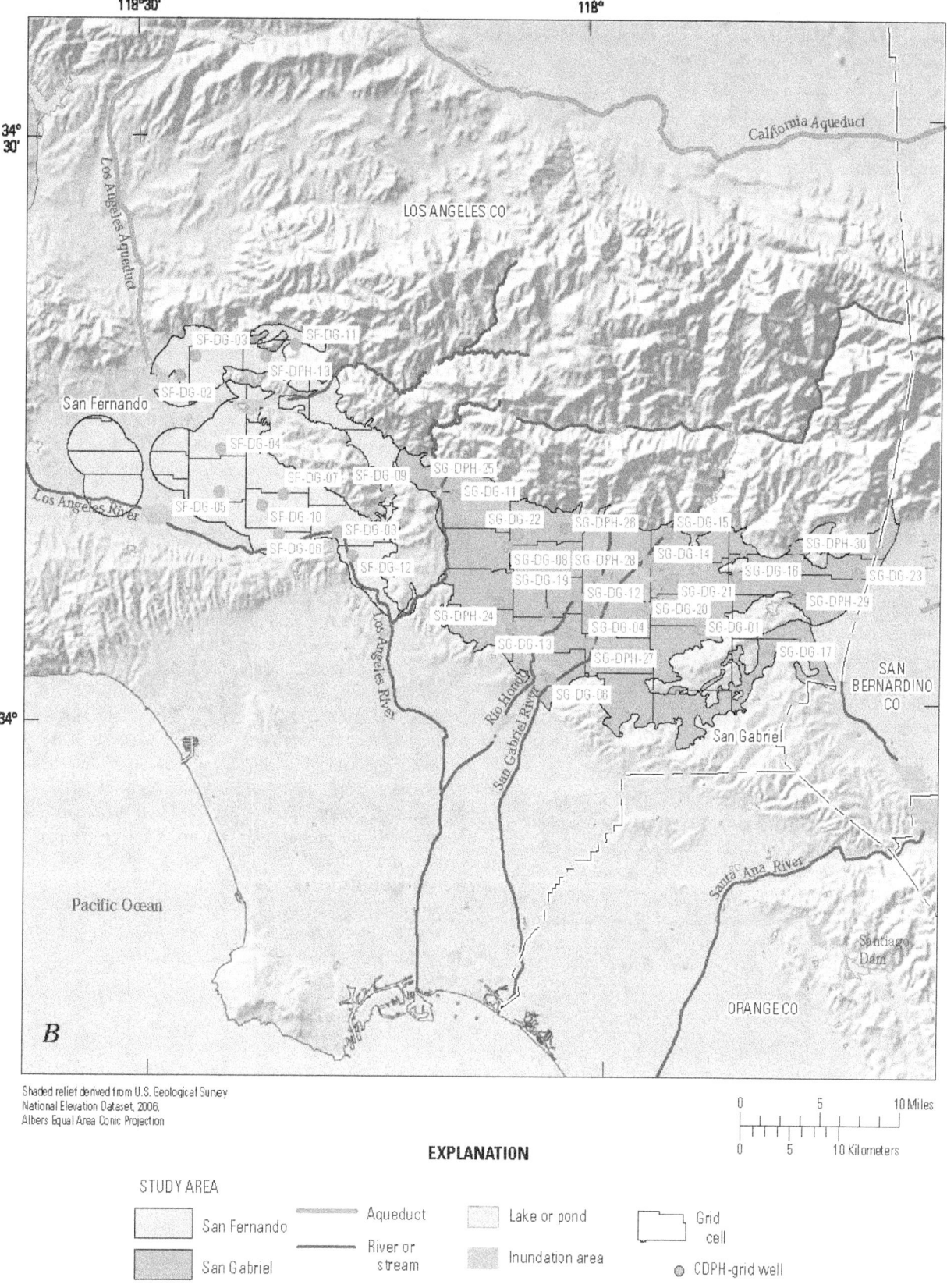

118°30'   118°

34°
30'

LOS ANGELES CO

California Aqueduct

SF-DG-03   SF-DG-11
SF-DPH-13
San Fernando   SF-DG-02
SF-DG-04
SF-DG-07   SF-DG-09   SG-DPH-25
SG-DG-11
Los Angeles River   SF-DG-05   SG-DG-22   SG-DPH-26   SG-DG-15   SG-DPH-30
SF-DG-10
SF-DG-06   SG-DG-08   SG-DPH-28   SG-DG-14   SG-DG-16   SG-DG-23
SF-DG-12   SG-DG-19   SG-DG-12   SG-DG-21   SG-DPH-29
SF-DPH-24   SG-DG-20
SG-DG-04   SG-DG-01
SG-DG-13   SG-DG-17
SG-DPH-27   SAN
BERNARDINO
CO
34°   SG-DG-06
San Gabriel

Pacific Ocean   Los Angeles River   Rio Hondo   San Gabriel River   Santa Ana River

Santiago
Dam
B   ORANGE CO

Shaded relief derived from U.S. Geological Survey
National Elevation Dataset, 2006,
Albers Equal Area Conic Projection

0   5   10 Miles

0   5   10 Kilometers

EXPLANATION

STUDY AREA

☐ San Fernando   ‒‒‒ Aqueduct   ☐ Lake or pond   ☐ Grid
                                                                    cell
☐ San Gabriel   ── River or   ☐ Inundation area   ⊙ CDPH-grid well
                            stream

Figure B1.—Continued

Appendix C. Estimation of Aquifer-Scale Proportions

Two statistical approaches, grid-based and spatially weighted, were selected to evaluate the aquifer-scale proportions of the primary aquifers in the FG study unit that had high, moderate, or low relative-concentrations (concentration relative to its water-quality benchmark) of constituents. The grid-based and spatially weighted estimations of aquifer-scale proportions, based on a spatially distributed grid cell network across the FG study unit, are intended to characterize the water quality of the primary aquifers, or at depths from which drinking water is usually drawn. These approaches assign weights to wells based on a single well per cell (grid-based) or the number of wells per cells (spatially weighted). Raw detection frequencies, derived from the percentage of the total number of wells with high or moderate relative-concentrations, also were calculated for individual constituents, but were not used for estimating aquifer-scale proportion because this method creates spatial bias towards regions with large numbers of wells.

1. Grid-based. One well in each grid cell, a "grid well," was randomly selected to represent the primary aquifers (Belitz and others, 2010). Most grid wells sampled for the FG study were USGS-grid wells. However, data for all constituents were not available for some USGS-grid wells, and additional data for CDPH-grid wells were selected to provide data for grid cells with no USGS-grid wells. The relative-concentration for each constituent (concentration relative to its benchmark) was then evaluated for each grid well. The proportion of the primary aquifers with high relative-concentrations was calculated by dividing the number of cells with concentrations greater than the benchmark (relative-concentration greater than 1) by the total number of grid wells in the FG study unit. Proportions containing moderate and low relative-concentrations were calculated similarly. Confidence intervals for grid-based aquifer proportions were computed using the Jeffreys interval for the binomial distribution (Brown and others, 2001). The

grid-based estimate is spatially unbiased. However, the grid-based approach may not identify constituents that exist at high concentrations in small proportions of the primary aquifers.

2. Spatially weighted. The spatially weighted approach relied on USGS-grid well data collected from May to August 2005 and CDPH data from May 1, 2002–April 30, 2005 (most recent analyses per well for all wells within each grid cell), and USGS-understanding public-supply well data. However, instead of data from only one well per grid cell, the spatially weighted approach uses all wells in each cell to calculate the high, moderate, and low relative-concentrations for the cell. The high, moderate, and low aquifer-scale proportions are then calculated from the percentage of cells with high, moderate, or low relative-concentrations (Isaaks and Srivastava, 1989). The resulting proportions are spatially unbiased (Isaaks and Srivastava, 1989). Confidence intervals for spatially weighted estimates of aquifer-scale proportion are not described in this report.

The raw detection frequency approach is the percentage (frequency) of wells within the study unit with high relative-concentrations. It was calculated by considering all of the available data collected during May 1, 2002–April 30, 2005, for the CDPH well data (the most recent analysis per well for all wells), the USGS-grid well data, and USGS-understanding wells. However, this approach is spatially biased because the USGS-understanding wells are not uniformly distributed (for example, figure 14K). Consequently, high values (or low values) for wells clustered in a particular area represent a small part of the primary aquifers, and could be given a disproportionately high (or low) weight compared to that given by spatially unbiased approaches. Raw detection frequencies of high relative-concentrations are provided to identify constituents for discussion in this report (table 4), but were not used to assess aquifer-scale proportions.

Appendix D. Comparison of California Department of Public Health and U.S. Geological Survey-GAMA Data

CDPH and USGS-GAMA data were compared to assess the validity of combining data from these different sources. Because laboratory reporting levels for most organic constituents and trace elements were substantially lower for USGS-GAMA data than for CDPH data (table 3), only relatively high concentrations of constituents could be compared, and as a result, there were insufficient data from which to evaluate agreement between CDPH and USGS-GAMA data. However, concentrations of inorganic constituents (sodium, calcium, fluoride, sulfate, TDS, and nitrate as nitrogen), which generally are prevalent at concentrations substantially greater than reporting levels, were compared for each well by using data from both sources. The USGS and CDPH databases contained data for major ions or the nutrient nitrate for 33 to 35 wells. Wilcoxon signed rank tests of paired analyses for these constituents indicated no significant differences between USGS-GAMA and CDPH data for these constituents. Although differences between the paired datasets occurred for some wells, most sample pairs plotted close to a 1:1 line (fig. D1). These plots indicated that the GAMA and CDPH inorganic data were comparable.

Major-ion data for grid wells with sufficient data (USGS and CDPH data) were plotted on a trilinear diagram (Piper, 1944) along with all CDPH major-ion data to determine whether the groundwater types in grid wells were similar to groundwater types observed historically in the study unit. Trilinear diagrams show the relative abundance of major cations and anions (on a charge equivalent basis) as a percentage of the total ion content of the water (fig. D2). Trilinear diagrams often are used to define groundwater type (Hem, 1970). All cation/anion data in the CDPH database with a cation/anion imbalance of less than 10 percent were retrieved and plotted on the trilinear diagram for comparison with USGS- and CDPH-grid well data.

The ranges of water types for USGS-grid wells and other wells from the historical CDPH database were similar (fig. D2). In most water samples from wells, no single cation accounted for more than 60 percent of the total cations, and bicarbonate accounted for more than 60 percent of the total anions. Waters in these wells are described as *mixed cation–bicarbonate* type waters. Many wells also contained *mixed cation–mixed anion* type waters, indicating that no single cation and no single anion accounted for more than 60 percent of the total.

The determination that the range of relative abundance of major cations and anions in grid wells is similar to the range of those in all CDPH wells indicates that the grid wells represent most of the types of water present in the FG study unit.

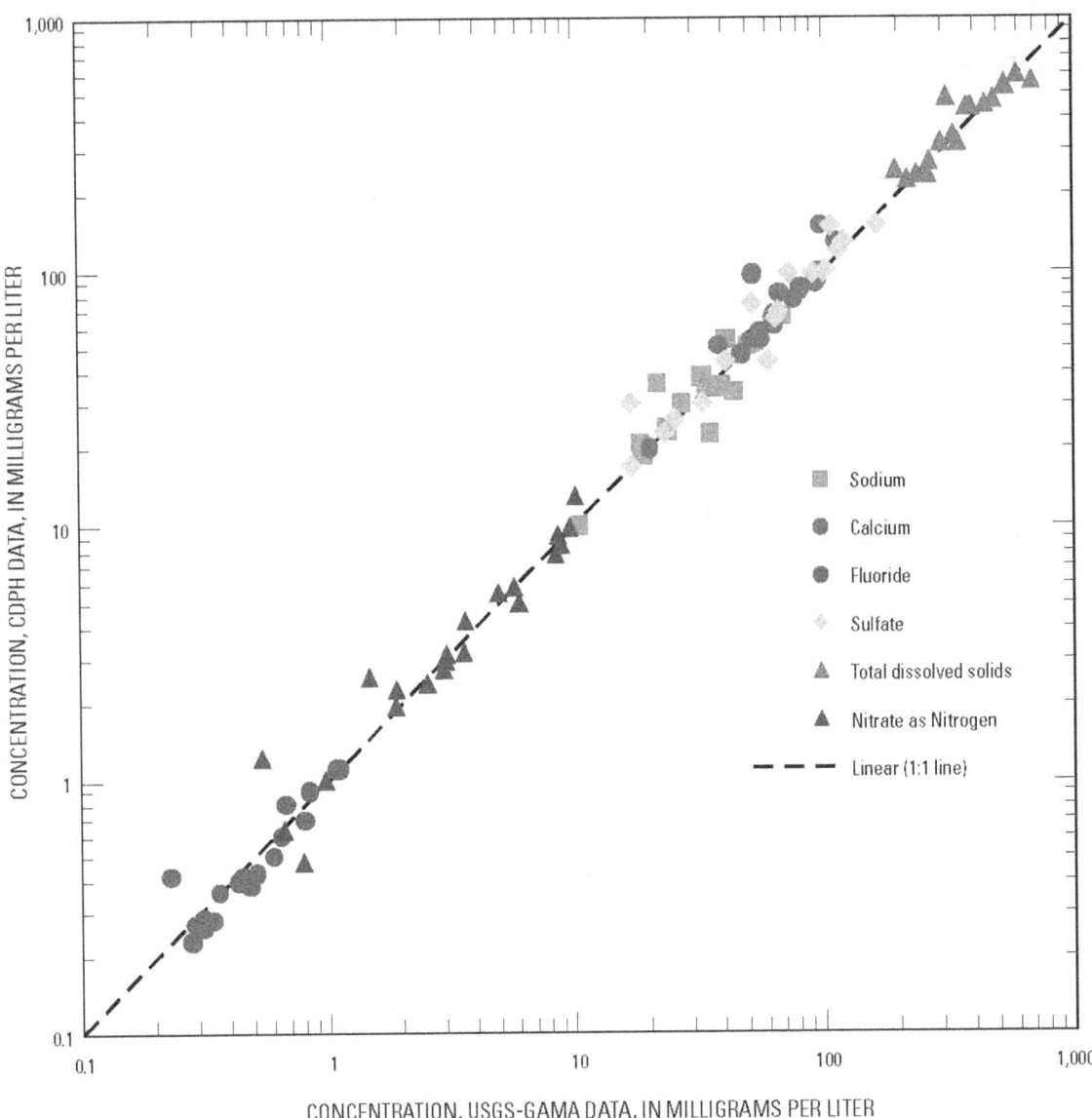

Figure D1.   Graph showing paired inorganic constituent concentrations from CDPH wells and wells sampled by the GAMA Program, May–August, 2005, San Fernando–San Gabriel study unit, California GAMA Priority Basin Project.

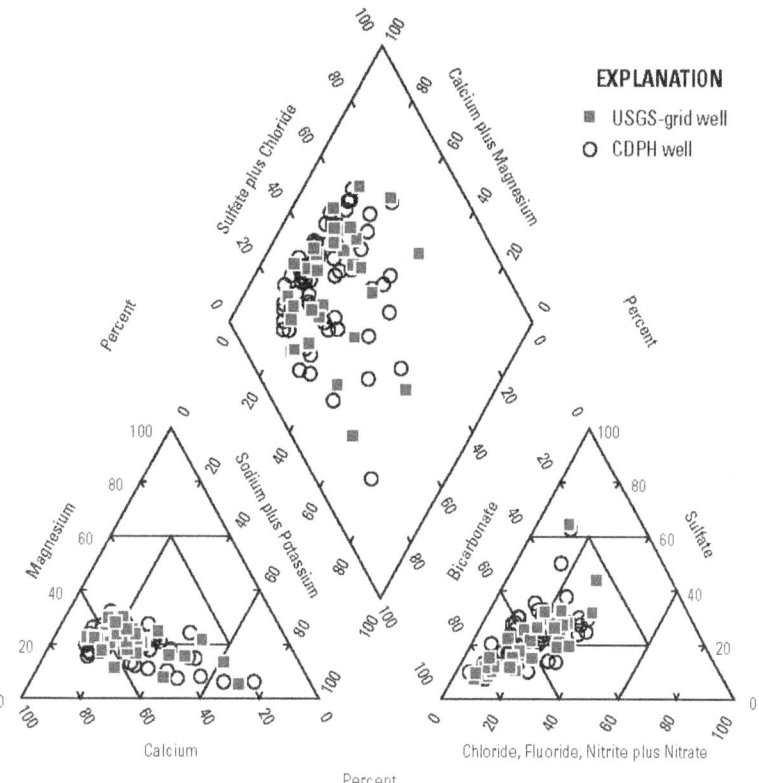

Figure D2. Trilinear diagram of selected inorganic data from USGS-grid wells and from all wells in the California Department of Public Health (CDPH) database that have a charge imbalance of less than 10 percent, San Fernando–San Gabriel study unit, California GAMA Priority Basin Project.